A Text Book Of

OPTICAL NETWORK AND SATELLITE COMMUNICATION

(22647)

Semester - VI

THIRD YEAR DIPLOMA IN ELECTRONICS ENGINEERING GROUPS

As Per MSBTE's 'I' Scheme Syllabus

Mrs. Pratibha D. Kulkarni
M.Tech., (E & TC)
Sr. Lecturer, E & TC Department
Pimpri-Chinchwad Education Trust's
Pimpri-Chinchwad Polytechnic.
Nigdi, PUNE - 44.

Miss Sharvari D. Kulkarni
M.E. (E & TC)

N4574

Optical Network and Satellite Communication   **ISBN 978-93-89825-03-9**

First Edition   :   **January 2020**

©   :   **Authors**

Published By :
NIRALI PRAKASHAN
Abhyudaya Pragati, 1312, Shivaji Nagar
Off J.M. Road, PUNE – 411005
Tel - (020) 25512336/37/39, Fax - (020) 25511379
Email : niralipune@pragationline.com

➢ DISTRIBUTION CENTRES

PUNE

Nirali Prakashan : 119, Budhwar Peth, Jogeshwari Mandir Lane, Pune 411002, Maharashtra
(For orders within Pune)   Tel : (020) 2445 2044, Fax : (020) 2445 1538; Mobile : 9657703145
   Email : niralilocal@pragationline.com

Nirali Prakashan : S. No. 28/27, Dhayari, Near Asian College Pune 411041
(For orders outside Pune)   Tel : (020) 24690204 Fax : (020) 24690316; Mobile : 9657703143
   Email : bookorder@pragationline.com

MUMBAI

Nirali Prakashan : 385, S.V.P. Road, Rasdhara Co-op. Hsg. Society Ltd.,
   Girgaum, Mumbai 400004, Maharashtra; Mobile : 9320129587
   Tel : (022) 2385 6339 / 2386 9976, Fax : (022) 2386 9976
   Email : niralimumbai@pragationline.com

➢ DISTRIBUTION BRANCHES

JALGAON

Nirali Prakashan : 34, V. V. Golani Market, Navi Peth, Jalgaon 425001, Maharashtra,
   Tel : (0257) 222 0395, Mob : 94234 91860; Email : niralijalgaon@pragationline.com

KOLHAPUR

Nirali Prakashan : New Mahadvar Road, Kedar Plaza, 1st Floor Opp. IDBI Bank, Kolhapur 416 012
   Maharashtra. Mob : 9850046155; Email : niralikolhapur@pragationline.com

NAGPUR

Nirali Prakashan : Above Maratha Mandir, Shop No. 3, First Floor,
   Rani Jhanshi Square, Sitabuldi, Nagpur 440012, Maharashtra
   Tel : (0712) 254 7129; Email : niralinagpur@pragationline.com

DELHI

Nirali Prakashan : 4593/15, Basement, Agarwal Lane, Ansari Road, Daryaganj
   Near Times of India Building, New Delhi 110002 Mob : 08505972553
   Email : niralidelhi@pragationline.com

BENGALURU

Nirali Prakashan : Maitri Ground Floor, Jaya Apartments, No. 99, 6th Cross, 6th Main,
   Malleswaram, Bengaluru 560003, Karnataka; Mob : 9449043034
   Email: niralibangalore@pragationline.com

Other Branches : Hyderabad, Chennai

niralipune@pragationline.com | www.pragationline.com

Also find us on 🅕 www.facebook.com/niralibooks

Dedication ...

Dedicated to my beloved Father

Shri. Krishnaji A. Deshpande (Principal)

Mother

Late Sau. Sunita K. Deshpande

– Mrs. Pratibha D. Kulkarni

Dedicated to my beloved Father

Shri. Dipak M. Kulkarni

Mother

Sr. Lecturer Sau. Pratibha D. Kulkarni

– Miss Sharvari D. Kulkarni

Preface ...

It gives me great pleasure and immense satisfaction in presenting this text book of **'Optical Network and Satellite Communication'** for students of Sixth Semester Diploma courses in Electronics Engineering Groups : DE/EJ/ET/EN/EX/EQ. This text has been written as per MBTE's 'I' scheme syllabus.

This subject is presented in a simple language considering the requirements of all the students regarding the changing trends of examination. Utmost care has been taken to check the mistakes and misprints, yet it is very difficult to claim perfection.

I take this opportunity to thank **Hon. Secretary Shri. V. S. Kalbhor, Hon. A. O. Shri. Vispute** and **Hon. Principal Mrs. V. S. Byakod, Vice-Principal Hon. Shri. B. V. Mane and Mr. Gorakh Bhalekar** for their moral support and encouragement.

I also thank **Shri. M. P. Munde** whose inspirational words and constant motivation helped me in completing this book in a short period. I am also thankful to the publisher **Shri. Dineshbhai Furia** and **Shri. Jignesh Furia.** I am also thankful to Mr. Malik Shaikh, Mrs. Anagha Kaware (Co-ordination and Proof Reading), Mrs. Yojana Deshpande (Figure Drawing) and staff of Nirali Prakashan for bringing out this book.

A very special thanks my father **Shri. Dipak Kulkarni, brother Chi. Shriprasad and husband Mr. Mangesh Gijare** for their support and co-operation.

Any undetected and unintentional error, omissions, suggestions etc. from students and teachers for improvement brought to our notice in good spirit are most welcome.

Pune.

Mrs. Pratibha D. Kulkarni
Miss Sharvari D. Kulkarni

Syllabus ...

1. Fundamentals of Fiber Optic Communication   (Marks 12, Hours 08)

1.1   Optical Fiber Communication : Advantages, Disadvantages, Applications.

1.2   Construction of Fiber Optic Cable.

1.3   Classification Based on Modes of Propagation of Light and Index Profile.

1.4   Optical Fiber Communication System : Block Diagram.

1.5   Optical Components : Sources and Detectors.

2. Optical Losses   (Marks 16, Hours 12)

2.1   Reflection, Refraction, Total Internal Reflection (TIR), Snell's Law, Critical Angle, Numerical aperture, Acceptance Angle and Acceptance Cone - (Numerical on above concepts).

2.2   Splicing Techniques - Fusion Splice, V-groove Splice and Elastic Tube Splice.

2.3   Losses in Optical Fiber: Absorption Loss, Scattering Loss, Dispersion Loss, Radiation Loss, Coupling Loss.

2.4   OTDR: Working Principle, Block Diagram, Specification, Application.

3. Optical Network   (Marks 14, Hours 08)

3.1   Optical Network Components Use and Features : Amplifiers, Splitter, Optical Switches.

3.2   WDM : Basic Concept, Features

3.3   SONET / SDH : Architecture and Hierarchy

3.4   Ethernet Standards of Optical Network Features : IEEE 802.3j, 802.3y, 802.3z

4. Overview of Satellite Systems   (Marks 16, Hours 12)

4.1   Working Principle, Concepts and Basic Components of Satellite System : Earth Segment, Space Segment, Active and Passive Satellite, Geostationary and Geosynchronous Satellites

4.2   Frequency Allocations for Satellite Services, Uplink and Downlink Frequency, Satellite Frequency Bands

4.3   Basic Terminologies used in Satellite Communication: Latitude, Longitude, Look Angle, Elevation Angle, Station Keeping, Propagation Delay Time Velocity, Look Angle and Footprint

4.4   Communication Satellite Orbits and its Types: LEO, MEO, Elliptical Orbit and GEO, Parameters and Characteristics of Various Orbits

4.5   Kepler's law, Apogee and Perigee Heights, Orbit Perturbations, Effects of a Non-spherical Earth, Atmospheric Drag, Effect of Eclipse on Satellite Motion

5. Satellite Segments and Services   (Marks 16, Hours 12)

5.1   Satellite Earth Station: Block Diagram; Antenna Subsystem, LNA, Power Subsystem, Telemetry Tracking and Command (TTAC) Subsystem, Attitude Control, Spinning Satellite Stabilization, Momentum Wheel Stabilization, Station Keeping, Thermal Control Transponder: Single, Double Conversion and Regenerative Type

5.2   Space Link: Equivalent Isotropic Radiated Power (EIRP), Transmission Losses : Free-space Transmission Loss, Feeder Losses, Antenna Misalignment Losses, Fixed Atmospheric and Ionosphere Losses

5.3   Satellite Applications:

    GPS: Global Positioning System (GPS) Concept, Working Principle, Transmitter and Receiver

    VSAT: Overview, Architecture, Working Principle, Applications

Contents ...

1. Fundamentals of Fiber Optic Communication   **1.1 - 1.20**

1.1	Introduction	1.1
1.2	Optical Spectrum	1.2
1.3	History of Fiber Optic	1.3
1.4	Advantages/Disadvantages of Fiber Optic Communication	1.4
1.5	Block Diagram of Fiber Optic Communication	1.5
1.6	Materials Used for Optical Fibres and Types	1.5
1.7	Applications of FOC	1.6
1.8	Construction of Fiber Optic Cable	1.6
1.9	Optical Fiber Types and Characteristics	1.6
	1.9.1 Multi-mode Step Index Fiber	1.8
	1.9.2 Multi-mode Graded Index Fiber	1.9
	1.9.3 Single-mode Step Index Fiber	1.10
	1.9.4 Comparison of Step Index and Graded Index Fiber	1.11
1.10	Optical Sources	1.11
1.11	LED (Light Emitting Diode)	1.11
	1.11.1 Edge Emitter LED	1.12
	1.11.2 Surface Emitting LED	1.13
1.12	Laser (Light Amplification by Stimulated Emission of Radiation)	1.13
1.13	Comparison of Different Sources	1.15
1.14	Optical Detectors	1.16
1.15	Photo Diodes	1.16
	1.15.1 PIN Diodes	1.16
	1.15.2 Avalanche Photo Diodes (APDs)	1.17
•	Important Points	1.18
•	Practice Questions	1.20

2. Optical Losses   **2.1 - 2.30**

2.1	Introduction	2.1
2.2	Definitions and Concepts	2.2
	2.2.1 Reflection	2.2
	2.2.2 Refraction	2.2
	2.2.3 Dispersion	2.3
	2.2.4 Diffraction	2.3
	2.2.5 Absorption	2.3
	2.2.6 Scattering	2.3
2.3	Propagation of Light (Energy) in Fiber Cable	2.3
	2.3.1 Snell's Law	2.4
	2.3.2 Critical angle of Fiber Optics	2.4
	2.3.3 Acceptance Angle of Optical Fiber	2.4
	2.3.4 Aumerical Aperture	2.5
2.4	Splicing Techniques	2.7
	2.4.1 Fusion Splice	2.8
	2.4.2 Elastic Tube Splice	2.9
	2.4.3 V-groove Splices	2.10
2.5	Fiber Connector	2.10
	2.5.1 Ferrule Connector	2.10
2.6	Losses in Optical Fiber	2.11
2.7	Absorption Loss	2.13
	2.7.1 Intrinsic Absorption	2.14
	2.7.2 Extrinsic Absorption	2.14

2.8	Scattering Loss	2.16
2.9	Linear Scattering Loss	2.16
	2.9.1 Rayleigh Scattering	2.16
	2.9.2 Mie Scattering	2.19
2.10	Non-linear Scattering	2.19
	2.10.1 Stimulated Brillouin Scattering	2.20
	2.10.2 Stimulated Raman Scattering	2.20
2.11	Dispersion Loss	2.21
	2.11.1 Intramodal Dispersion	2.22
	2.11.2 Material Dispersion	2.22
	2.11.3 Waveguide Dispersion	2.23
	2.11.4 Intermodal Dispersion	2.23
2.12	Radiation Loss	2.24
2.13	Coupling Loss	2.24
	2.13.1 Lateral Misalignment (Displacement)	2.25
	2.13.2 Gap Misalignment (Displacement)	2.25
	2.13.3 Angular Misalignment (Displacement)	2.25
	2.13.4 Imperfect Surface Finish	2.25
2.14	Optical Time Domain Reflectometry (OTDR)	2.26
•	Important Points	2.27
•	Practice Questions	2.29

3. Optical Network — **3.1 - 3.10**

3.1	Introduction	3.1
3.2	Optical Network Elements	3.1
	3.2.1 Amplifiers (Optical Amplifiers)	3.2
	3.2.2 Optical Splitters	3.3
	3.2.3 Optical Switches	3.4
3.3	Wavelength Division Multiplexing (WDM)	3.6
	3.3.1 Features of WDM	3.6
	3.3.2 Advantages of WDM	3.7
	3.3.3 Disadvantages of WDM	3.7
3.4	SONET / SDH	3.7
	3.4.1 Advantages of SONET	3.8
	3.4.2 Disadvantages of SONET	3.8
3.5	Ethernet Standard	3.8
•	Important Points	3.10
•	Practice Questions	3.10

4. Overview of Satellite Systems — **4.1 - 4.14**

4.1	Introduction to Satellite	4.1
4.2	Importance of Satellite Communication System	4.2
4.3	Concept of Orbit and its Types	4.3
4.4	Various Components of Satellite System	4.4
4.5	Communication Link - Uplink and Downlink Frequency	4.6
	4.5.1 Satellite Speed and Period	4.8

4.6	Look Angles	4.8
	4.6.1 Elevation Angle	4.8
	4.6.2 Azimuth Angle	4.9
4.7	Altitude	4.9
4.8	Foot-print	4.10
4.9	Frequency Bands used in Satellite Communication	4.10
4.10	Kepler's Law	4.11
	4.10.1 Kepler's First Law	4.11
	4.10.2 Kepler's Second Law	4.11
	4.10.3 Kepler's Third Law	4.11
	4.10.4 Apogee and Perigee Heights	4.12
4.11	Orbit Perturbations	4.12
	4.11.1 Effects of a Non-spherical Earth	4.12
	4.11.2 Atmospheric Drag	4.12
•	Important Points	4.13
•	Practice Questions	4.14

5. Satellite Segments and Services **5.1 - 5.12**

5.1	Introduction	5.1
5.2	Satellite Earth station block diagram	5.1
5.3	Satellite Subsystems	5.2
	5.3.1 Power Subsystem	5.3
	5.3.2 Communication Channel Subsystem	5.4
	5.3.3 Attitude Control Subsystem	5.4
	5.3.4 Attitude Control Subsystem	5.4
	5.3.5 Telemetry Tracking and Command Subsystems	5.5
	5.3.6 Main and Auxiliary Propulsion Subsystem	5.5
	5.3.7 Antenna System	5.5
	5.3.8 The Satellite Transponder	5.6
	5.3.9 Sation Keeping	5.7
5.4	Space Link	5.8
	5.4.1 Equivalent Isotropic Radiated Power (EIRP)	5.8
5.5	Transmission Losses	5.8
	5.5.1 Free-Splace Transmission Loss	5.9
	5.5.2 Feeder Loss	5.9
	5.5.3 Antenna Misalignment Loss	5.9
	5.5.4 Atmospheric and Ionospheric Losses	5.9
5.6	Satellite Applications	5.9
	5.6.1 Global Positioning System (GPS)	5.10
	5.6.2 Very Small Aperture Terminals (VSATs)	5.11
•	Important Points	5.11
•	Practice Questions	5.12

***

Chapter 1

Fundamentals of Fiber Optic Communication

Weightage of Marks = 12, Teaching Hours = 08

Syllabus

1.1 Optical Fiber Communication : Advantages, Disadvantages, Applications.
1.2 Construction of Fiber Optic Cable.
1.3 Classification Based on Modes of Propagation of Light and Index Profile.
1.4 Optical Fiber Communication System : Block Diagram.
1.5 Optical Components : Sources and Detectors.

About this Chapter

At the end of this chapter students will be able to:
- Describe construction and features of optical fiber.
- Compare working of optical fiber for given mode and index profile.
- Explain the block diagram of optical fiber communication system.
- Explain the working principle of given optical source and detector.

1.1 INTRODUCTION

- The communication medium in most electronic communication systems is either a wire conductor cable or free space.
- One of the main limitations of communication systems is their restricted information carrying capabilities.
- As we know, the information handling ability is directly proportional to the bandwidth of the communications channel.

  For example, in telephonic systems, the bandwidth is limited by the characteristics of the cable used to carry the signals.
- As the demand for telephones has increased, better cables and wiring systems have been developed.
- Further, multiplexing techniques have been developed to transmit multiple telephone conversations over a single cable.
- These techniques have the same effect as the number of cables or channels of communications were greatly multiplied.
- But now-a-days, a new medium is growing in popularity that is fiber optic cable.
- A fiber optic cable is a glass pipe used to carry a light beam from one place to another.
- Light is an electromagnetic signal like a radio wave.
- It can be modulated by information and sent over the fiber optic cable.
- Because the frequency of light is extremely high, it can accommodate very wide bandwidths of information and extremely high data rates can be achieved with excellent reliability.

1.2 OPTICAL SPECTRUM (S-16; W-16)

- Light is a kind of electromagnetic radiation.

- Another familiar form of electromagnetic radiation is a radio wave.

- Any electromagnetic wave is made-up of both electric and magnetic fields that travel through space from one place to another place.

- The basic important characteristics of electromagnetic radiation is its frequency or wavelength.

- To put light waves into perspective with the lower radio and other frequencies shown in Fig. 1.1.

(W-15; S-16)

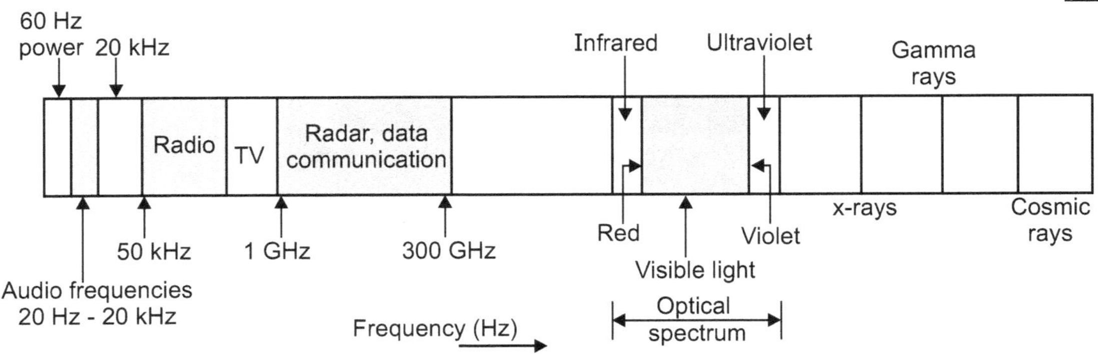

Fig. 1.1: Electromagnetic Frequency Spectrum

- At left end of the spectrum are very low frequency electrical signals such as 60 Hz ac power and audio frequencies in the range of 20 Hz to 20 kHz.

- These are low frequencies that have long wavelengths.

- Higher up the scale to the right side is a wide range of radio frequencies.

- Radio frequency range is from 10 kHz to 300 GHz (approx.).

- Microwaves extend from 1 GHz to 300 GHz.

- Even further up the scale we find visible light.

- **This visible light range between infrared and ultraviolet rays is called optical spectrum, of the range $3 \times 10"$ to 3×10^6 Hz.**

- This includes both infrared and ultraviolet as well as visible parts of the spectrum.

- The visible spectrum ranges from 4.3×10^{14} to 7.5×10^{14} Hz, but we rarely refer to the frequency of light.

- Instead, we state light in terms of wavelength.

  (Where, wavelength is a distance measured in meters between peaks of a wave).

- Light waves are very short and are usually expressed in nanometers or micrometers or microns.

- Visible light is in the range of 400 to 700 nanometers (nm) or 0.4 to 0.7 µm (micrometers), depending on the colour of light.

- Short wavelength light is violet (400 nm) and real (700 nm) is long-wavelength light.

- Right below visible light is a region known as infrared.

  (Infrared rays cannot be seen).

- However, they have the properties of light and therefore, can be manipulated in the same way with mirrors, lenses and other devices.

- The source of most infrared is heat.

- Right above the visible light is the ultraviolet range.

- Infrared and ultraviolet are included in optical spectrum.

1.3 HISTORY OF FIBER OPTIC

- One way to expand communication capability further is to use light as the transmission medium.

- Instead of using an electrical signal travelling over cable or electromagnetic waves travelling through space, the information is put on a light beam and transmitted through space or through a special cable.

- The use of visible optical carrier waves or light for communication has been for many years.

- Simple systems such as signal fires, reflecting mirrors and more recently, signaling lamps have provided successful if limited information transfer.

- Moreover, as early as 1880 **Alexander Graham Bell,** the inventor of the telephone, demonstrated that information could be transmitted by light.

- A special transmitter made-up of a microphone attached to a mirror was used to transmit voice over a distance of several hundred feet using a light beam from the sun.

- However, although some investigation of optical communication continued in the early part of the twentieth century and its use was limited to mobile, low capacity communication links.

- Communication by light beam in free space is impractical over very long distances because of great attenuation of the light due to atmospheric effects like fog, haze, smog, rain, snow and other conditions absorb, reflect and refract the light, greatly attenuating it and thereby limiting the transmission distance.

- Artificial light beams used to carry information are also virtually obliterated during day light hours by the sun.

- Light beam communications was made more practical in the early 1960's with the invention of the **laser.**

- The laser is special height-intensity, signal frequency light source.

- It produces a very narrow beam of brilliant light of specific wavelength (i.e. colours).

- Because of its great intensity, the laser beam can penetrate atmospheric obstacles better than other types of light, thereby making light beam communication more reliable over long distances.

- The primary problem with such free-space light beam communication is that the transmitter and receiver must be perfectly aligned.

- Instead of using free space, some type of light carrying cable can also be used.

- For centuries, it has been known that light is easily transmitted through various types of transparent media such as glass and water.

- By the mid-1950's, glass fibers were developed that permitted long light-carrying cables to be constructed.

- Over the years these glass fibers have been perfected.

- Further, low-cost plastic fiber cable was also developed.

- Today, these fiber optic cables have been highly refined.

- Cables may miles long can be constructed and interconnected for the purpose of transmitting information on a light beam over very long distances.

- These highly refined fiber optic cables, a new transmission medium is now available.

- Its great advantage is that the light beam have an incredible information capacity.

- Whereas hundreds of telephone conversations may be transmitted simultaneously at microwave frequencies thousands of signals can be carried on a light beam through a fiber optic cable.

- In parallel with the development of the fiber waveguide attention was also focused on the other optical components.

- Since optical frequencies are accompanied by extremely small wavelengths the development of all these optical components required a new technology.

- Thus, semiconductor optical sources (Laser and LED) and detectors (photodiodes and phototransistors) compatible in size with optical fibers were designed and fabricated to enable successful implementation of the optical fiber system.

- Initially, the semiconductor lasers exhibited very short life times of at best a few hours, but significant advances in the device structure enabled life times greater than 1000 hr and 700 hr to be obtained by 1973 and 1977 respectively.

- These devices was originally fabricated from alloys of (gallium arsenides (AlGaAs) which emitted in the near infrared between 0.8 and 0.9 µm.

- To obtain both low loss and low dispersion at the same operating wavelength, new advanced single mode fiber structures have been realized, namely, dispersion shifted and dispersion flattened fibers.

- Hence, developments in fiber technology have continued rapidly over recent years.

- High performance, reliable optical fiber communication systems are now widely deployed both within telecommunications networks and many others more localized communication application areas.

1.4 ADVANTAGES/DISADVANTAGES OF FIBER OPTIC COMMUNICATION

(S-15; W-15, 16)

QUESTION
1. State the advantages of optical fiber communication (any four points).   **(4M)**

The advantages of optical fiber cable over electrical cable are as follows –

(1) Greater bandwidth:

As carrier frequency is higher, the bandwidth is 106 times greater.

(2) Low loss:

Loss in co-axial cable and twisted pair wire increases with frequency whereas loss in the optical cable remains flat over a very wide range of frequencies. i.e. loss in optical fiber is slow.

(3) Electromagnetic immunity:

Glass fiber neither pick-up nor generate Electromagnetic Interference (EMI) like co-axial cables, because glass is an insulator and optical fiber has cladding. (outer coverings).

(4) Light weight:

A glass fiber weights less than a copper conductor because the copper requires more lines than fiber.

A coaxial cable weights nine times greater than fiber. (In aircraft and automobiles).

(5) Small size:

A single fiber is capable of replacing a very large bundle of individual copper wires. For example, in telephone cable, 1000 pairs of copper wire equals the single fiber optical cable capable of handling same amount of signal.

(6) Safety:

A fiber is a dielectric. It does not carry electricity. It presents no spark or fire hazards, it does not attract lightening. It is possible to run fiber directly through a fuel tank.

(7) Security:

Fiber optic is a highly secure transmission medium. It does not radiate energy that can be received by a nearby antenna, and it is difficult to tap a fiber.

- **Disadvantages:**

  There are some disadvantages of fiber optic cable.

  1. Its small size and brittleness make its difficult to work with.
  2. Special expensive tools and techniques required.
  3. High cost.

1.5 BLOCK DIAGRAM OF FIBER OPTIC COMMUNICATION (S-16)

QUESTION

1. Explain the block diagram of fiber optic communication   (8M)

- A block schematic of general fiber optic communication system is shown in Fig. 1.2.

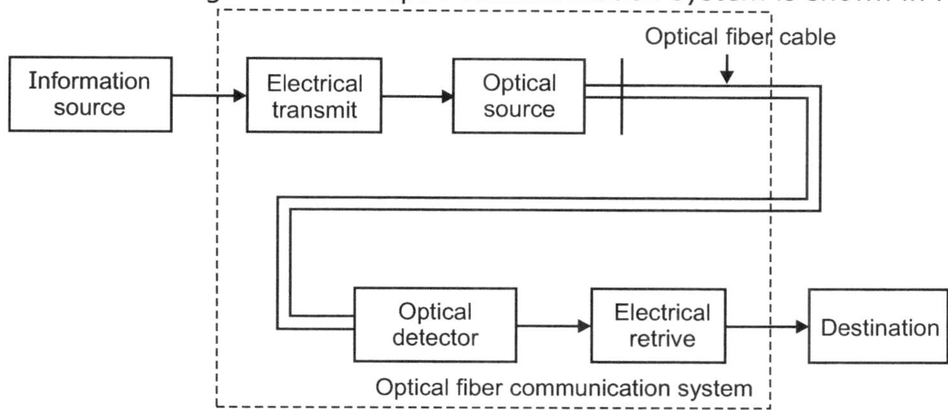

Fig. 1.2: Optical Fiber Communication System (W-15)

Functions of Each Block:

Information Source:

  The original information such as sound, data, picture is converted into electrical signal.

Electrical Transmit:

  It drives the optical source.

Optical Source:

  Converts electrical signal into light using laser or LED.

Optical Fiber Cable:

  It is transmission media to travel light beam.

Optical Detector:

  At receiver side optical detector converts light beam into an electrical signal.

Electrical Retrive:

  Provides demodulation of optical carrier.

Destination:

  Again it is original information.

1.6 MATERIALS USED FOR OPTICAL FIBRES AND TYPES

  Two types of optical fiber available as regards to material used.

  (1) Glass Fibers, (2) Plastic Fibers.

(1) Glass Fiber:

  Glass is not a particular material, but it is a super cooled liquid because glass is a solid state of matter in which atoms are not in regular array.

- When a liquid is cooled down very rapidly, then at temperature below the freezing temperature, the liquid gets solidify in glassy or non-crystalline state as shown in Fig. 1.3.

- The material used for making fibers is silicon dioxide (silica).

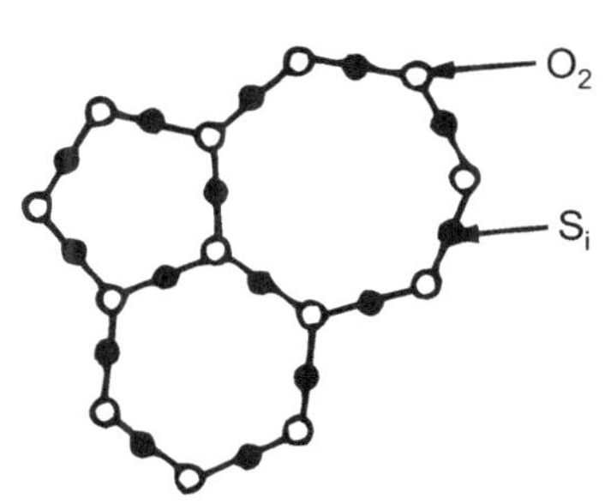

Fig. 1.3

- The silicon dioxide is mixed with oxides of genermanium boron, phosphorous to adjust its optical density slightly from that of silica. Because there are different layers of glass having different refractive indices.

(2) Plastic Fiber:

- The plastic material used for making fibers is polystyrene or poly-vinyl chloride.
- These fibers has higher losses and low withstanding capacity for temperatures.

1.7 APPLICATIONS OF FOC (FIBER OPTIC COMMUNICATION)

QUESTIONS

1. Illustrate in detail any two applications of optical fiber communication. **(4M)**
2. State any four applications of FOC. **(4M)**

1. Local and long distance telephone systems.
2. TV studio-to-transmitter interconnection, eliminating microwave radio link.
3. Closed-circuit TV systems used in buildings for security.
4. Secure communications systems at military bases.
5. Shipboard communications.
6. Computer networks, wide area and local area.
7. Aircraft controls.
8. Aircraft communications.
9. Data acquisition and control signal communications in industrial process control system.
10. Nuclear plant instrumentation.
11. Utilities (electric, gas etc.) station communications.
12. College campus communications.

1.8 CONSTRUCTION OF FIBER OPTIC CABLE (W-15)

QUESTIONS

1. Describe construction of optical fiber. **(4M)**
2. Draw the constructional sketch of fiber optic cable and give its classification. **(4M)**

- The fiber, which is called the core, is usually surrounded by a protective cladding shown in Fig. 1.4.
- The cladding is also made of glass or plastic but has lower index of refraction.
- This ensures that proper interface is achieved so that light waves remain within the core.
- In addition to protecting the fiber core from nicks and scratches, the cladding adds strength.

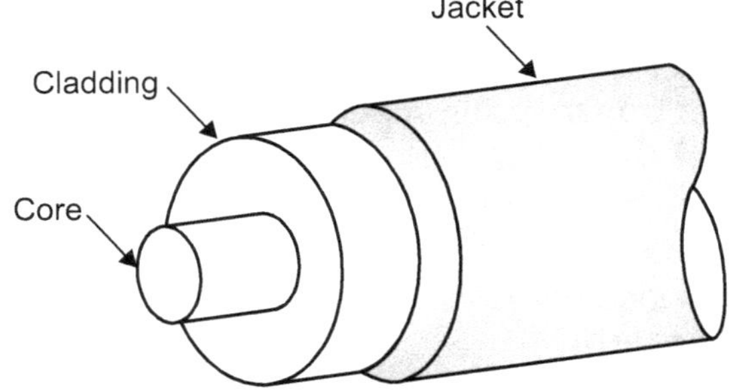

Fig. 1.4: Basic Construction of Fiber Optic Cable

  [Some fiber-optic cables have a glass core with a plastic cladding].
- Others have a plastic core with a plastic cladding.
- Another common arrangement is a glass core with a plastic cladding. It is called plastic clad silica (PCS) cable.
- A plastic jacket as a insulation is put over the clad.

1.9 OPTICAL FIBER TYPES AND CHARACTERISTICS

QUESTIONS

1. Give classification of optical fiber. **(4M)**
2. How light signal propagates through optical fiber? **(4M)**

The characteristics of light transmission through a glass fiber depends on many factors such as:

1. The composition of the fiber.
2. The amount and type of light introduced into the fiber.
3. The diameter and length of the fiber.

- The composition of the fiber determines the refractive index.
- By a process called doping, in which other materials are introduced into the material that change its index number. This process produces a single fiber with a core index n_1 and a surface index (cladding) n_2 typically $n_1 = 1.48$ and $n_2 = 1.46$.

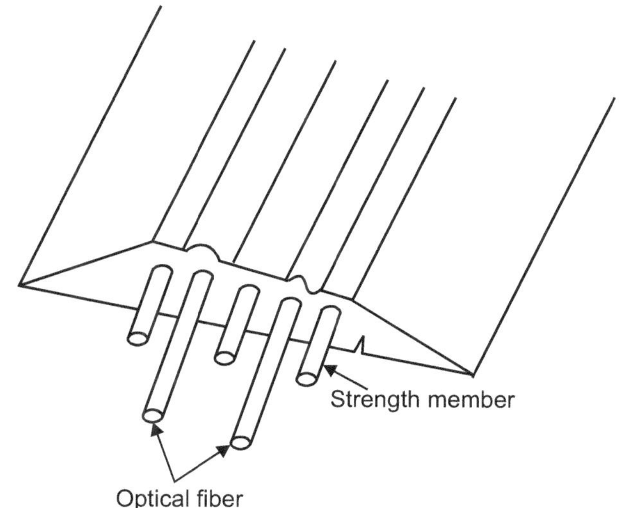

Fig. 1.5: Fiber Cable Assembly

- Another characteristics of the fiber, which depends on its size, is its **mode of operation.**
- The term **mode** is used refers to mathematical and physical descriptions of the propagation of energy through a medium.
- The number of modes supported by a single fiber can be as low as 1 or as high as 1,00,000 so that fiber can provide a path for one light ray or for hundreds of thousands of light rays.
- From this characteristics introduce the terms **single mode** and **multimode.**
- Another term which is the refractive index profile.
- It describes the relationship between the multiple indices which exist in the core and cladding of the particular fiber i.e. **"light changes speed when it passes from one medium to another".**
- There are two major indices:
1. Step index.
2. Graded index.
- **Classification:**

The **Refractive Index profile** describes the relation between the indices of core and cladding.

Mode: Mode is a mathematical and physical concept desiring the propagation of electromagnetic wave through optical fiber.

A mode is simply a path that a light ray can follow in travelling down a fiber. The number of modes supported by a fiber ranges from one to over 1,00,000. Thus, a fiber provides a path of travel for one or thousands of light rays, depending on its size and properties.

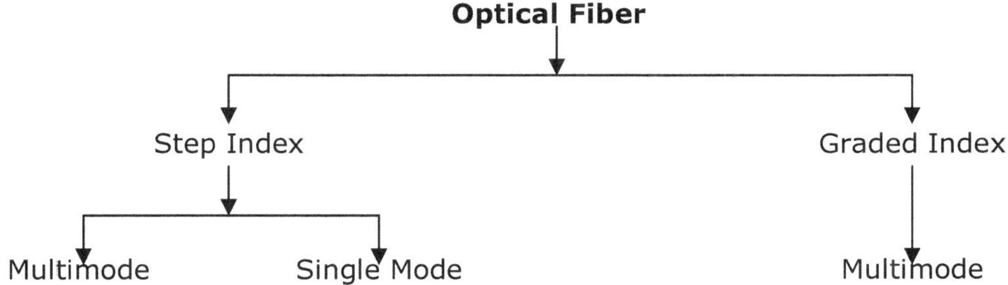

Considering the refractive index profile and mode optical fiber is classified as –
1. Multi-mode Step Index Fiber.
2. Multi-mode Graded Index Fiber.
3. Single-mode Step Index Fiber.

- The step index describes an abrupt index change from the core to the cladding.
  For example, a core with a uniform index (1.48) and a cladding with a uniform index. (1.46).
- In graded index fiber, the heighest index is at the center (1.48). This number decreases gradually until it reaches the index number of the cladding (1.46) i.e. near the surface.

1.9.1 Multi-mode Step Index Fiber

- It is also known as simply Step-Index Fiber.
- **The fiber which has uniform distribution of refractive index n_1 throughout its core is called as Step Index Fiber.**
- The cladding has uniform refractive index distribution as shown in Fig. 1.6

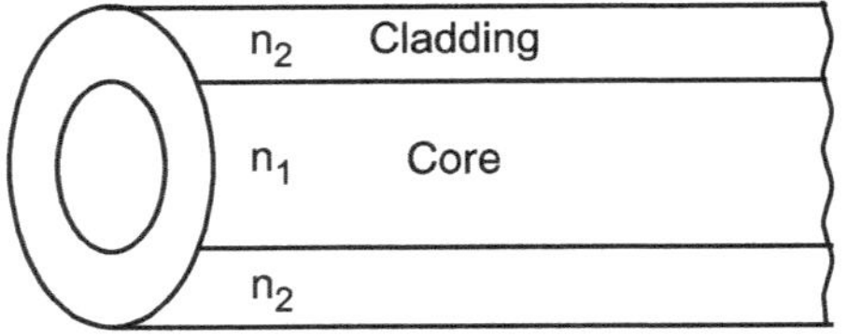

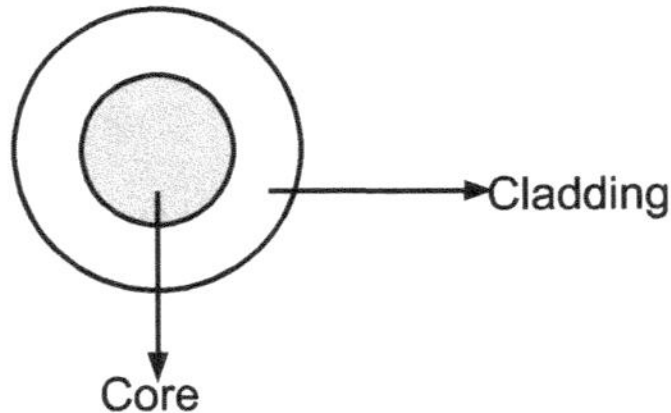

Fig. 1.6: Uniform Distribution of Refractive Index in Step Index Fiber

- Step index refers to rapid change in refractive index at the core-clad interface.
- The refractive indices of core and cladding are slightly different.
- Refractive index of core is more than that of cladding.
- From core to clad, refractive index immediately changes in step i.e. from n_1 to n_2.
- **Multimode Propagation** in Step Index consists of many light rays.
- Step Index provides reflection of light rays because of immediate change in refractive index between the core and cladding. It gives zigzag transmission of light.

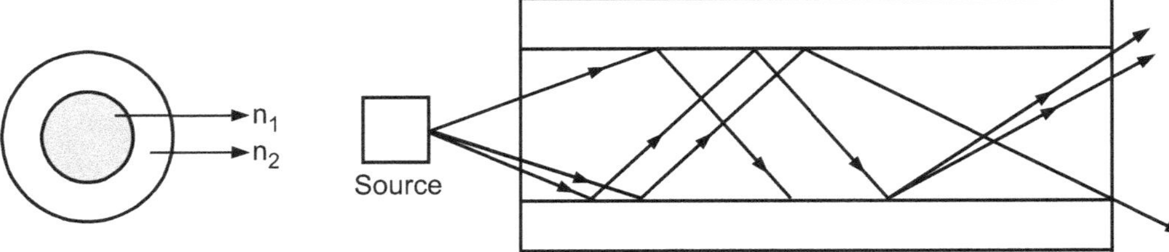

Fig. 1.7: Multimode propagation in Step-Index Fiber

Advantages:

1. Multimode step-index fiber is of large size.
2. Typical core diameter are in 50 to 1000 µm range.
   Such large diameter cores are excellent of gathering light and transmitting it efficiently i.e. an inexpensive light source such as LED can be used to produce the light pulses.
3. It is also easiest to make.

Application:

Multimode-step-index is widely used for short to medium distances at relatively low pulse frequencies.

Limitation:

- In Fig. 1.8, a short light pulse is applied to the end of cable by the source.
- Light rays from the source will travel by multiple path.
- At the end of cable, those rays which travel the shortest distance reach the end first.
- Other rays begin to reach the end of the cable later in time until the light ray with the longest path finally reaches the end concluding the pulse.
- Ray A reaches end first, then B and then C.

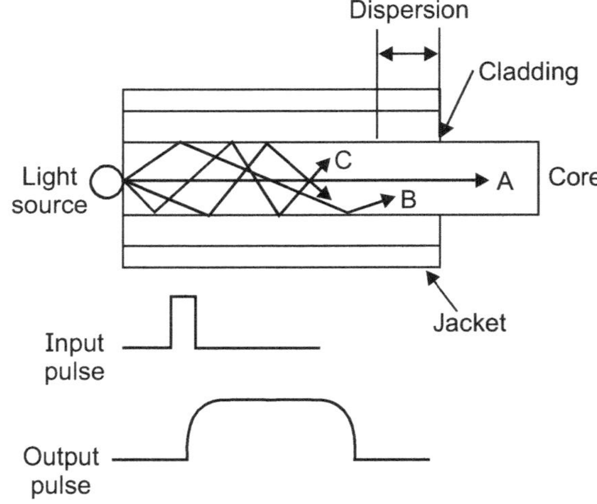

Fig. 1.8: A Multimode Step Index

- The result is a pulse at the other end of the cable that is lower in amplitude due to the attenuation of the light in the cable and increased duration due to the different arrival times of various light rays. **This stretching of the pulse is known as modal dispersion.**

- But the pulse has been stretched, input pulses cannot occur at a rate faster than the output pulse duration permits.

- Otherwise the pulses will merge shown in Fig. 1.9

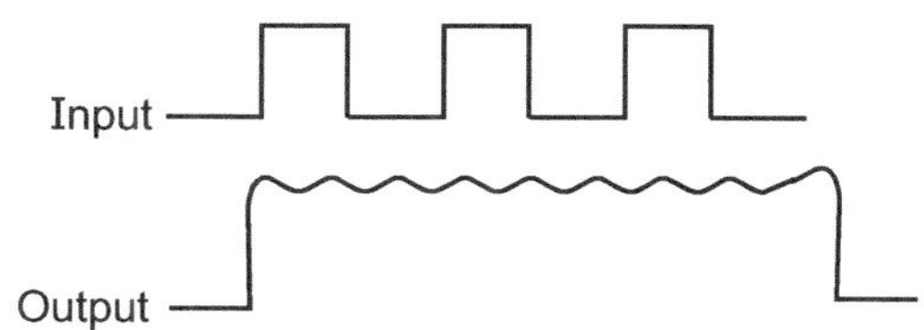

Fig. 1.9: The Effect of Modal Dispersion on Pulses occurring too rapidly in a Multimode Step Index Cable

- At the output, one long pulse occurs and will be indistinguishable from the three separate pulses originally transmitted.

- i.e. incorrect information will be received.

Solution to above problem:

- The only cure for this problem is to reduce the repetition rate or the frequency of the pulses.

- When this is done, proper operation occurs.

- But with pulses at lower frequency, less information can be handled.

1.9.2 Multi-mode Graded Index Fiber

- It is also known as simply Graded Index Fiber. It has a non-uniform core.

- **The core index is highest at center and decreases until it matches that of cladding.**

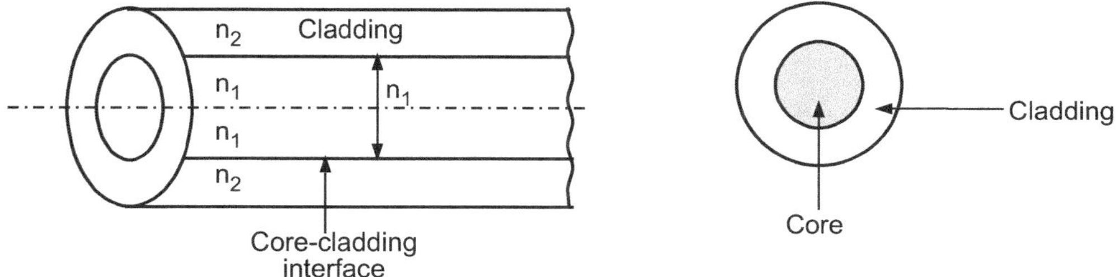

Fig. 1.10: Non-uniform Core in Graded Index Fiber

- Here the refractive index of core represented as n_1.

- So that value of n_1 is maximum at center of the fiber i.e. at the axis.

- Away from center, value of refractive index decreases.

- But this minimum value of n_1, at the core-clad interface is always greater than refractive index of cladding.

Multi-mode Propagation

- In graded index, more gradual change of refractive index occurs, hence light rays are bent towards the axis of fiber in the form of refraction.

- It gives gradual bending effects of light as shown in Fig. 1.11

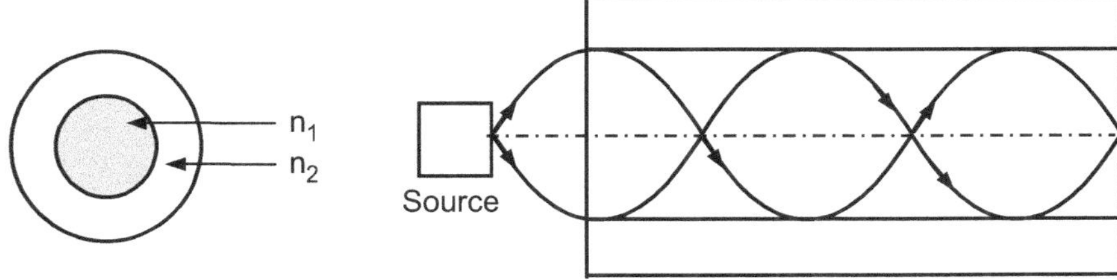

Fig. 1.11: Multi-mode Propagation in Graded Index Fiber

Advantage over Multimode-Step-Index:

- Fig. 1.12 shows the typical paths of the light beams.

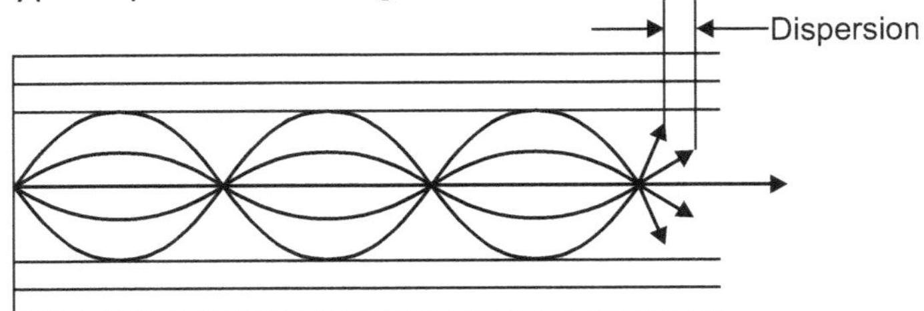

Fig. 1.12: Dispersion in Multimode Graded-Index Cable

- Because of continuously varying refractive index across the core, the light ray bent smoothly and coverge rapidly at points along the cable.
- The light ray near the edge of the core take a longer path but travel faster since refractive index is less.
- All light paths tend to arrive at one point simultaneously.
- This results in less modal dispersion.
- It is not eliminated entirely, but the output pulse is not nearly as stretched as in multimode step-index cable.
- The output pulse is only slightly elongated.
- **As a result, this cable can be used at very high pulse rates and thus considerable amount of information can be carried on it.**

Advantages:

- This type of cable is also much wider in diameter with core size in the **50 to 100 µm range.**
    1. Therefore, it is easier to splice and interconnect.
    2. Cheaper and less-intense light sources may be used.

Application:

Used for very high pulse rates.

1.9.3 Single-mode Step Index Fiber

- A single mode fiber guides the light ray in straight line and has core of smaller diameter as compared to multi-mode fiber.
- In single mode (mono) light travels along single path.
- It is used for extremely high data rate applications like long distance telephone transmission. Here core has smaller diameter as shown in Fig. 1.13.

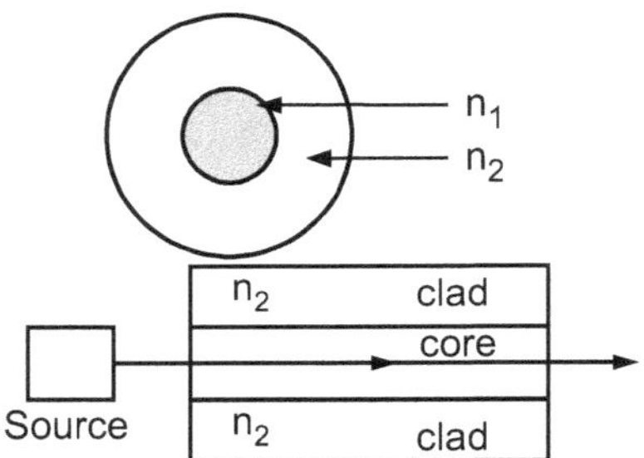

Fig. 1.13: Single Mode Step Index Fiber

- Due to small core, the total number of paths through the core are minimized and modal dispersion is eliminated.
- Typical core size are 2 to 15 µm.
- With minimum refraction no stretching occurs.
- The output pulse has essentially the same duration as the input pulse.

Advantages:

The single mode step index fibers are best since the pulse repetition rate can be high and the maximum amount of information can be carried.

Disadvantages:

1. The main problem with this type of cable is because of very small size, it is difficult to make and is therefore very expensive.
2. Handling, splicing and making interconnections are more difficult.
3. Requires superintense light source such as laser.

Application:

For long distances single mode step index cable is used.

1.9.4 Comparison of Step Index and Graded Index Fiber

Sr. No.	Parameter	Step-Index Fiber	Graded-Index Fiber
1.	Data rate	Slow	High
2.	Coupling efficiency	High	Low
3.	Index variation	$\Delta = \dfrac{n_1 - n_2}{n_1}$	$\Delta = \dfrac{n_1^2 - n_2^2}{2n_1^2}$
4.	Ray path	By total internal reflection	Travels in oscillatory manner
5.	Numerical aperature	NA remain same.	Changes continuously with distance from fiber axis.
6.	Material	Plastic or glass	Only glass
7.	Attenuation	Less 0.34 dB/km	More 0.6-1 dB/km
8.	Light source	LED	LED, LASER
9.	Application	Subscriber local network communication	Local network and WAN

1.10 OPTICAL SOURCES

- In an optical communication systems, transmission begins with the transmitter, which consists of a modulator and the circuitary that generates the carrier.
- In optical communication, carrier is a light beam that is modulated by digital pulses which turn in ON and OFF.
- **The basic transmitter is known as source.**
- **Conventional light sources such as incandescent lamps cannot be used in fiber optic systems** because they are simply too slow.
- An incandescent light source consists of a filament that heat up and emits light.
- Such a light source cannot be turned OFF and ON fast enough because of the thermal delay in the filament.
- So that, in order to transmit high-speed digital pulses, a very fast light source must be used.
- The two most commonly used light (optical) sources are:
  1. LEDs and　　　　　　　　　　2. LASERs.

1.11 LED (LIGHT EMITTING DIODE)

QUESTION

1. State four limitations of LED as a source to optical fiber.　　　　　**(4M)**

- An LED is a p-n junction diode usually made from a semiconductor material such as aluminium-gallium-arsenide (AlGaAs) or gallium-arsenide-phosphide (GaAsP).

Working Principle:

- LEDs emit light by spontaneous emission, light emitted as a result of the recombination of electrons and holes.
- When current is passes through the PN structure of LED, electrons are injected into an active layer. Light is generated when these injected electrons return from their excited (higher) energy levels to their normal (lower) energy levels. The energy difference is releases as light.

Construction:

- The construction of the LED is shown in Fig. 1.14.

- In the Fig. 1.14, the surface layer is p–type.

- A shallow p-n junction is formed and electrical contacts made to both regions.

- The upper part of the p-material is left uncovered so that radiations from the device are not much affected.

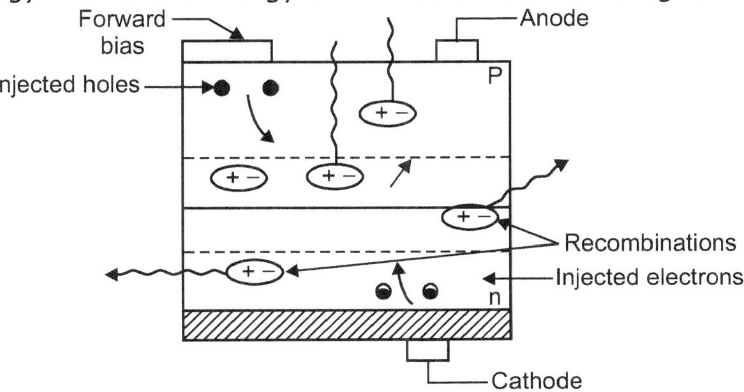

Fig. 1.14: LED Construction

- The inner quantum efficiency of the LED material, sometimes is nearly 100%.
- The external efficiency is much lower.
- This is so since most of the emitted radiation strikes the material interface at an angle which is greater than the critical angle and hence, gets trapped.
- The semiconductor/air interface in the LED has the shape of a hemisphere.
- This is made so such that most of the rays strike the surface at less than the critical angle.

Working:

- When forward biased, minority carriers are injected across the P-N junction.
- Once across the junction, these minority carriers recombine with majority carriers and give-up energy in the form of light.

Types of LED:

1. Surface emitter LED.
2. Edge emitter LED.
3. Super luminescent LED.
4. Planar LED.
5. Dome LED.

Advantages:

As already stated, LED has a number of advantages. They have –

(a) very high speed (in n sec)
(b) different colours
(c) long life
(d) low voltage operation
(e) low cost
(f) linearity of radiant power output with forward current over a wide range.
(g) low voltage operation to make it compatible with ICs.

Disadvantages:

However, it has some outstanding disadvantages.

(a) It is sensitive to over voltage and over current due to which it is easily damaged.
(b) The radiant output power and wavelength are temperature dependent.
(c) It is difficult to get the theoretical overall efficiency except by using special cooled or pulse conditions.
(d) Large power dissipation.

Applications:

1. They are used in 7-segment, 16-segment and dot-matrix displays, which are used to indicate alphanumeric characters and symbols.
2. They are used for indicating power ON/OFF conditions.
3. They are used in burglar alarm systems.
4. They are used to indicate digital logic state.
5. They can be used as solid state video displays in place of CRTs.
6. They are used in image sensing circuits in picture phones.

Colours:

- By using different LED materials and different dopants, it is possible to get different colours.
- LEDs may be made from gallium phosphide (GaP), gallium arsenide (GaAs) and silicon carbide (SiC). SiC gives yellow light, gallium arsenide phosphide gives red light while gallium arsenide gives infrared radiation. GaP with nitrogen gives green light.

1.11.1 Edge Emitter LED 

- The edge emitting LED which was developed by RCA, shown in Fig. 1.15.
- These LEDs emit a more directional light pattern than surface emitting LED.

Construction:

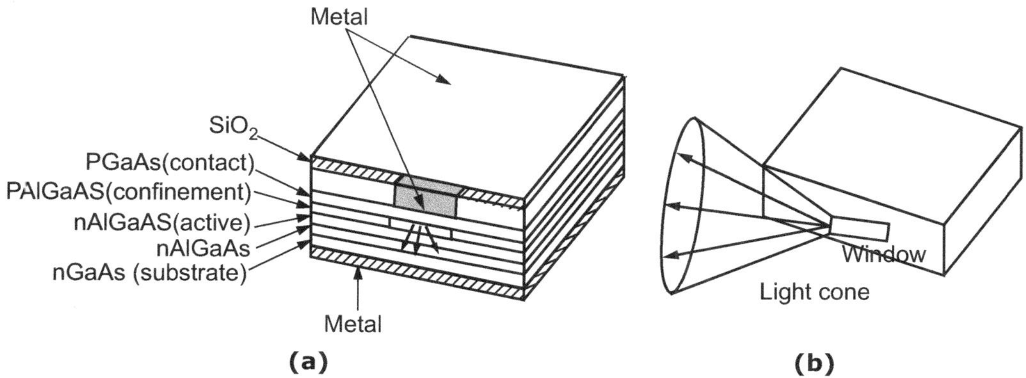

Fig. 1.15: Edge Emitting LED

- The construction is similar to the planar and surface emitting diodes except that the emitting surface is stripe rather than a confined circular area.
- The light is emitted from an active stripe and forms an elliptical beam.
- The radiant light power emitted from an LED is a linear function of the forward current passing through the device.
- It can also be seen that the optical output power of an LED is, in part, a function of the operating temperature.

1.11.2 Surface Emitting LED

- For more practical applications, such as telecommunications, data rates in excess of 100 Mps are required. For these applications etched-well surface emitting LED was developed.

Construction:

- Fig. 1.15 (c) shows surface emitting LED.
- This LED emits light in many directions.
- The etched well helps to concentrate the emitted light to a very small area.
- These devices are more efficient than the standard surface emitters, and they allow more power to be coupled into the optical fiber, but they are also more difficult to be coupled into the optical fiber.

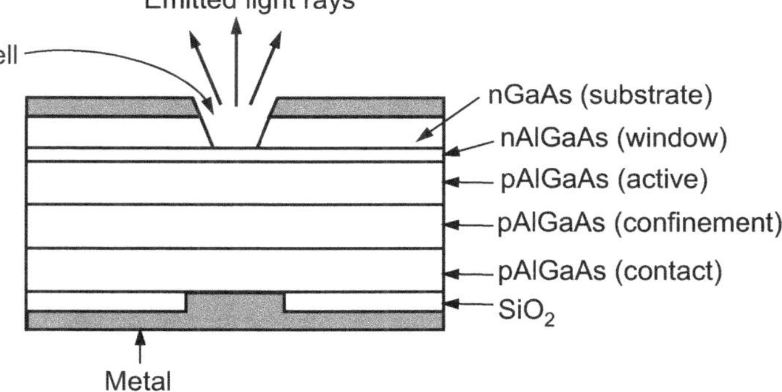

Fig. 1.15 (c): Surface Emitting LED

1.12 LASER (LIGHT AMPLIFICATION BY STIMULATED EMISSION OF RADIATION)

- The laser is a device which amplifies the light, hence the LASER is an acronym for light amplification by stimulated emission of radiation.
- The operation of the device may be described by the formation of an electromagnetic standing wave within a cavity (optical resonator) which provides an output of monochromatic highly coherent radiation.

Principle:

- **Laser action is the result of three process absorption of energy packets (photons), spontaneous emission, and stimulated emission.**
- **These processes are represented by the simple two-energy-level diagrams.**

    where,   E1 is the lower state energy level.

               E2 is the higher state energy level.

- Quantum theory states that any atom exists only in certain discrete energy state, absorption or emission of light causes them to make a transition from one state to another.
- The frequency of the absorbed or emitted radiation is related to the difference in energy E between the two states.

  If   E1 is lower state energy level

  and   E2 is higher state energy level

  $$E = E2 - E1 = h \cdot f$$

  where, $h = 6.626 \times 10^{-34}$ Js (Planck's constant).

- An atom is initially in the lower energy state, when photon with energy E2 - E1 is incident on the atom it will be excited into the higher energy state E2 through the absorption of the photon.

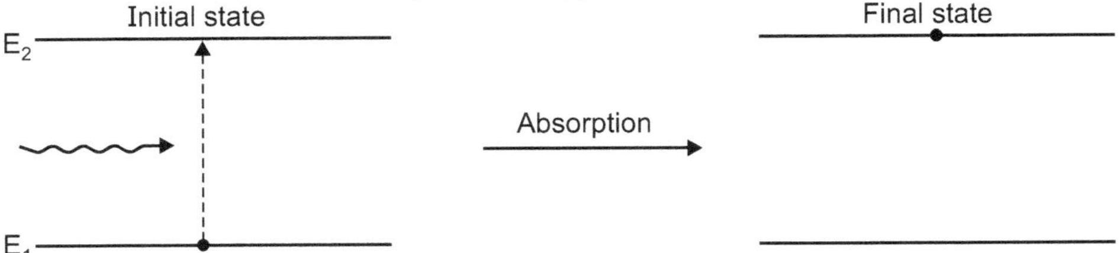

Fig. 1.16: Absorption

- When the atom is initially in the higher energy state E2, it can make a transition to the lower energy state E1 providing the emission of a photon at a frequency corresponding to $E = h \cdot f$.
- The emission process can occur in two ways:

  (a) By spontaneous emission in which the atom returns to the lower energy state in random manner.

  (b) By stimulated emission when a photon having equal energy to the difference between the two states (E2 - E1) interacts with the atom causing it to the lower state the creation of the second photon.

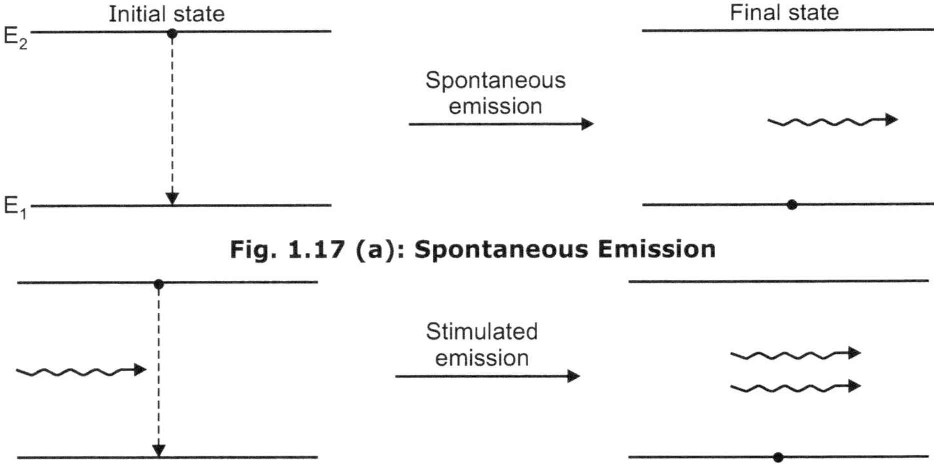

Fig. 1.17 (a): Spontaneous Emission

Fig. 1.17 (b): Stimulated Emission

- Spontaneous emission gives incoherent radiation.
- Stimulated emission gives coherent radiation.
- Since the light associated with emitted photon is of same frequency of incident photon, and in same phase with same polarization.
- It means that when an atom is stimulated to emit light energy by an incident wave, the liberated energy can add to the wave in constructive manner.
- The emitted light is bounced back and forth internally between two reflecting surfaces.
- The bouncing back and forth of light wave causes their intensity to reinforce and build-up.
- The result is a high brilliance, single frequency light beam providing amplification.

Construction:

- LASERs are constructed for many different materials, including gases, liquids and solids.

- The construction of LASER is similar to LED except that the ends are highly polished.

- The mirror like ends trap the photons in the active region and as they reflect back and forth, stimulate free electrons to recombine with holes at higher-than normal energy level. This is called lasing.

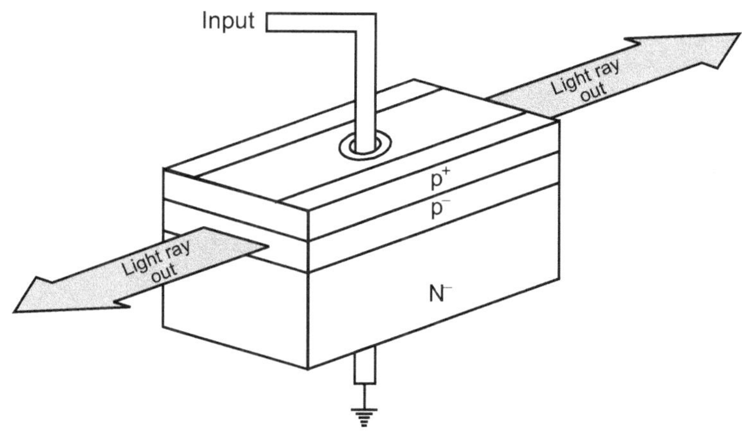

Fig. 1.18: LASER Diode Construction

Working:

- As current passes through a forward biased P-N junction diode, light is emitted by spontaneous emission at a frequency determined by the energy gap of the semiconductor material.

- When a particular current level is reached the number of minority carrier and photons produced on either side of the P-N junction reaches a level where they begin to collide with already excited minority carriers.

- This causes an increase in the ionization energy level and makes the carriers unstable.

- When this happens, a typical carrier recombine with an opposite type of carrier at an energy level that is above its normal before collision value.

- In the process, two photons are created, one is stimulated by another.

- Essentially, a gain in the number of photons is realized.

- For this to happen, a large forward current that can provide many carriers (holes and electrons) is required.

Advantages over LED:

1. Developing light power upto several watts.
2. More powerful than LEDs & therefore capable of transmitting much lower distances.
3. Ability to turn OFF and ON at a faster rate.
4. High speed laser diodes capable of gigabit per second digital data rates.

1.13 COMPARISON OF DIFFERENT SOURCES   (S-16)

QUESTION

1. Compare between LED and LASER.   **(4M)**

Sr. No.	Parameter	LED	LASER
1.	Principle of operation	Spontaneous emission	Stimulated emission
2.	Spectral width	Broad (20-100 nm)	Narrow (1-5 nm)
3.	Data rate	Low	Very high
4.	Transmission distance	Smaller	Greater
5.	Temperature sensitivity	Less	More
6.	Coupling efficiency	Very low	High
7.	Compatible fibers	Multi-mode step index	Single mode step index
8.	Cost	Low	High
9.	Noise	More	Less

1.14 OPTICAL DETECTORS

QUESTION

1. State four requirements of optical detector. Describe working principle of photodiodes.   

Predominant types of receiver used in communication system rely on principle of ionization of semiconductor.

Optical receivers are placed at the receiver side to convert light ray into electrical signal.

Four important parameters:

1. Detector responsivity:

This is the ratio of output current to input optical power. Hence, this is efficiency of the device.

2. Spectral response range:

This is the range of wavelength over which device will operate.

3. Response time:

This is the time required for responding variation in light intensity.

4. Noise:

Level of noise produced during lowest level of light detection.

- **Requirements of Detectors:**
  1. The detector should be highly sensitive at operating wavelength.
  2. Produce a maximum electrical signal strength for a given amount of optical power.
  3. Noise introduced by detector should be minimum.
  4. Stable performance characteristics.
  5. Small physical size.
  6. Less bias voltage or current requirement.
  7. Less cost.

1.15 PHOTO DIODES

QUESTIONS

1. What is the use of photo diode in fiber optic communication ? Explain its working principle. List its limitations.   **(4M)**
2. Describe operation, principle of PIN diode.   **(4M)**

- Photo diodes convert light directly to electric current.
- PIN diode can convert one photon to one electron of current. Hence, amplifier is needed with receiver.

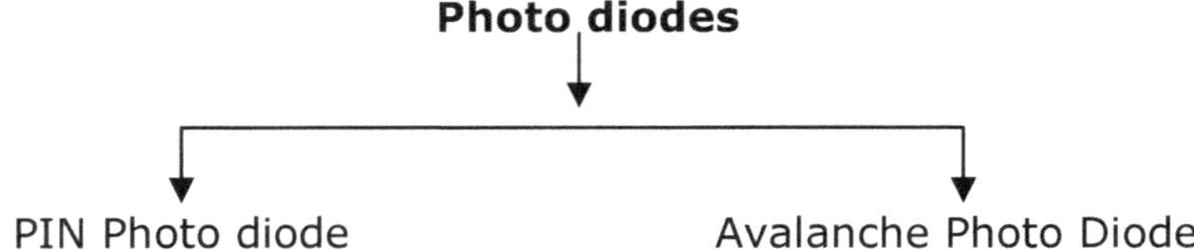

1.15.1 PIN Diodes   **(S-15; W-15)**

Name PIN because intrinsic layer introduced between P and N region.

Principle:

Principle of PIN diodes is reverse that of LED. That is, light is absorbed at p-n junction rather than emitted.

The wide intrinsic i layer has only small degree of doping and has wide depletion layer. There are number of modifications:

1. It increases the chances of entering photon being absorbed, because amount of absorbant material is more.
2. Since junctions are wider, junction capacitance is less.
3. Two ways of current flow are diffusion and drift.

Thus i layer increases responsivity and decreases response time.

Construction:

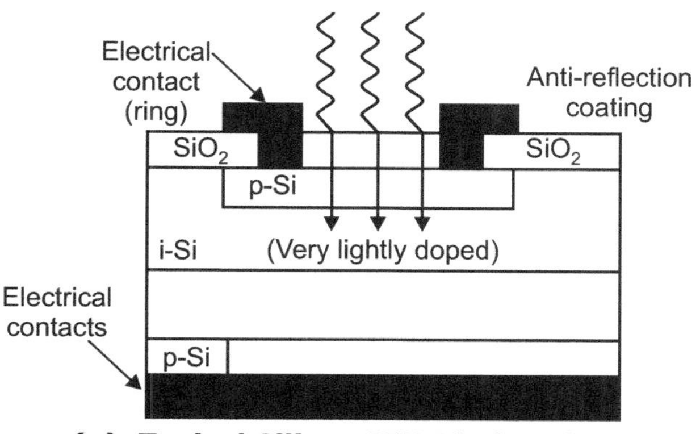

(a) Typical Silicon PIN Diode Schematic   **(b) How diode converts light into Voltage Pulses**

Fig. 1.19

Working:

- The sensitivity and response time can be increased by adding an undoped intrinsic (I) layer between the P and N semiconductor to form a PIN diode.
- The I layer exposed to light.
- This is a silicon PN junction diode that is sensitive to light.
- This diode is normally reverse-biased.
- The only current that flows through it is an extremely small leakage current.
- Whenever light strikes the diode, this leakage current will increase significantly.
- It flows through a resistor and develops a voltage drop across it.
- The result is an output voltage pulse.
- Voltage pulse is small, so that it must be amplified.
- For operating PIN diode, energy of absorbed photon must be sufficient to promote an electron across bandgap.
- At energy higher than bandgap energy, material absorbs photons.
- Germanium is material used as it has low bandgap energy regardless of wavelength.

Characteristics of PIN diodes:

1. Quantum efficiency:

It is defined as the ratio of number of electrons collected at the junction over the number of incident photons. PIN diodes has 1 electron release per 1 photon.

2. Responsivity:

It is a measure that does take photon energies into account. It is output photo current of device divided by the input optical power.

Typical responsivity for $\lambda = 900$ nm is 0.44.

1.15.2 Avalanche Photo Diodes (APDs)   `(S-16; W-15, 16)`

- APDs amplify the signal during the detection process. They use a similar principle to that of photomultiplier tubes used in nuclear detection.

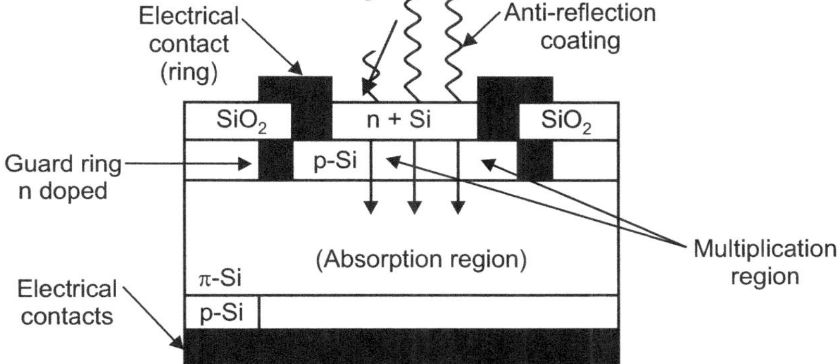

Fig. 1.20: Avalanche Photo Diode (APD)

Principle:

A single photon acting on the device releases a single electron. This electron is accelerated through an electric field until it strikes a target material.

This collision releases multiple electrons. These electrons get accelerated and strike another target.

This releases more electrons. This process is repeated. Thus, one electron has created large current.

Working:

- Structural difference between APD and p-i-n diode is that i zone (which is lightly n-doped in p-i-n structure) is lightly p-doped and renamed π layer.
- Arriving photons generally pass straight through n^+-p junction (because it is thin) and are absorbed in π layer.
- This absorption produces free electron.
- Electric potential across π layer is sufficient to attract the electron towards one contact and holes at other contact.
- Electrons are attracted towards n^+ layer at top of the device.
- Around the junction between n^+ and p-layers the electric field is so intense that carriers are strongly accelerated.
- When these electrons collide with other atom, they produce electron-hole pair.
- This process is called as impact ionization.
- When the light strikes the junction, breakdown occurs and large current flows.
- This high reverse current requires less amplification.
- Like PIN diode APD also reverse biased.
- When a sufficient amount of reverse will flow due to avalanche effect.

APD Characteristics:

1. Sensitivity of APDs:

Very high sensitivity makes APD very useful.

2. Gain bandwidth product:

Performance of optical detector is decided by its gain bandwidth product.

APD has gain bandwidth product of 150 GHz.

3. Operating speed:

Device capacitance limit speed of p-i-n diodes.

APD takes too long to make impact ionization effect.

Hence, maximum speed of APD is restricted.

Advantages:

- It is fastest and most sensitive photodiode.
- Capable to handle very high gigabit/sec. data rates possible in some systems.

Disadvantage:

- It is expensive and circuitry is complex.

Important Points

- An **optical fiber** is like electrical transmission line in which light is transmitted.
- The **name optical fiber** because the signal in the form of light travels inside the glass fiber due to internal reflection.
- Because of the frequency of light is extremely high it can accommodate very wide bandwidths of information and extremely high data rates can be achieved with excellent reliability.

- Light is a kind of electromagnet radiation.
- The visible light range with infrared and ultraviolet rays is called **optical spectrum.**
- Basic elements in fiber optic system are:

  1. Information source. 2. Optical source. 3. Optical detector.
- A fiber optic cable is a glass pipe used to carry a light beam from one place to another.
- Light is a kind of electromagnetic radiation.
- This visible light range between infrared and ultraviolet rays is called optical spectrum, of the range $3 \times 10"$ to 3×106 Hz.
- Thus, semiconductor optical sources (Laser and LED) and detectors (photodiodes and phototransistors) compatible in size with optical fibers were designed and fabricated to enable successful implementation of the optical fiber system.
- A single fiber can handle as many voice channels as a 1500 pair cable can.
- Fiber is immune to interference from lightning, cross-talk and electromagnetic radiation.
- The characteristics of light transmission through a glass fiber depends on many factors such as:

  1. The composition of the fiber.
  2. The amount and type of light introduced into the fiber.
  3. The diameter and length of the fiber.
- There are two major indices:

  1. Step index.
  2. Graded index.
- The **Refractive Index profile** describes the relation between the indices of core and cladding.
- **Mode:** Mode is a mathematical and physical concept desiring the propagation of electromagnetic wave through optical fiber.
- The fiber which has uniform distribution of refractive index n1 throughout its core is called as Step Index Fiber.
- **Multimode Propagation** in Step Index consists of many light rays.
- Multimode-step-index is widely used for short to medium distances at relatively low pulse frequencies.
- The core index is highest at center and decreases until it matches that of cladding.
- As a result, this cable can be used at very high pulse rates and thus considerable amount of information can be carried on it.
- A single mode fiber guides the light ray in straight line and has core of smaller diameter as compared to multi-mode fiber.
- **The basic transmitter is known as source.**
- **Conventional light sources such as incandescent lamps cannot be used in fiber optic systems** because they are simply too slow.
- An LED is a p-n junction diode usually made from a semiconductor material such as aluminium-gallium-arsenide (AlGaAs) or gallium-arsenide-phosphide (GaAsP).
- LEDs emit light by spontaneous emission, light emitted as a result of the recombination of electrons and holes.
- **Principle:** Laser action is the result of three process absorption of energy packets (photons), spontaneous emission, and stimulated emission. (These processes are represented by the simple two-energy-level diagrams).
- Predominant types of receiver used in communication system rely on principle of ionization of semiconductor.

- PIN diode can convert one photon to one electron of current. Hence, amplifier is needed with receiver.

Photo diodes

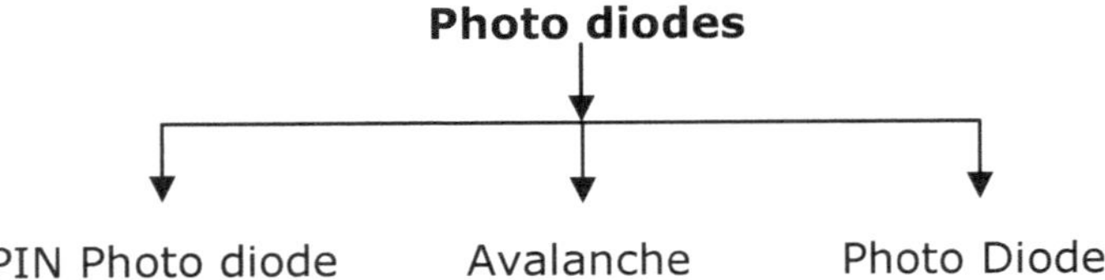

PIN Photo diode   Avalanche   Photo Diode

- **Principle:** Principle of PIN diodes is reverse that of LED. That is, light is absorbed at p-n junction rather than emitted.
- APDs amplify the signal during the detection process. They use a similar principle to that of photomultiplier tubes used in nuclear detection.

Practice Questions

1. Draw and explain light wave spectrum.
2. Draw the block diagram of fiber optics and explain.
3. State the advantages of fiber optics
4. State applications of FOC.
5. Give optical sources and detectors.
6. Explain LED and LASER with neat diagram of each.
7. Compare step index and graded index fiber.
8. Give the classification of optical fiber.
9. Explain multimode step index fiber.
10. Explain multimode graded index fiber.
11. Explain single mode step index fiber.
12. Draw the structure of edge limiting LED and explain.
13. Compare LED and LASER.
14. Give classification of optical detectors.
15. Explain PIN photo diode with neat diagram.
16. Draw avalanch photo diode and explain.
17. Give advantages and disadvantages of APD.

Chapter 2

Optical Losses

Weightage of Marks = 16, Teaching Hours = 12

Syllabus

2.1　Reflection, Refraction, Total Internal Reflection (TIR), Snell's Law, Critical Angle, Numerical aperture, Acceptance Angle and Acceptance Cone - (Numerical on above concepts).

2.2　Splicing Techniques - Fusion Splice, V-groove Splice and Elastic Tube Splice.

2.3　Losses in Optical Fiber: Absorption Loss, Scattering Loss, Dispersion Loss, Radiation Loss, Coupling Loss.

2.4　OTDR: Working Principle, Block Diagram, Specification, Application.

About this Chapter

At the end of this chapter students will be able to:

- Explain the given terms related to optical theory.
- Calculate acceptance angle, critical angle and numerical aperture of the given optical fiber cable.
- Explain the step-by-step procedure of given splicing techniques.
- Describe the different types of Optical fiber losses.
- Explain the operation of ODTR.

2.1 INTRODUCTION

- Because of rapidly increasing demands for telephone communication throughout the world, multiconductor copper cables have become not only very expensive but also insufficient way to meet these information requirements.

- The frequency limitations in the copper conductor system approximately 1 MHz.

- The optical fiber with its low weight and high frequency characteristics approximately 40 GHz and its imperviousness to interference from electromagnetic radiation, have become the choice for all heavy demand long-line telephone communication systems.

Why to use Optical Fibers ?

Reasons are:

1. The light weight and non-corrosiveness of fiber make it very practical for aircraft and automotive applications.

2. A single fiber can handle as many voice channels as a 1500 pair cable can.

3. The spacing of repeaters from 35 to 80 km for fibers as opposed to from 1 to $1\frac{1}{2}$ km for wire is great advantage.

4. Fiber is immune to interference from lightning, cross-talk and electromagnetic radiation.

- In order to plan the use of optical fibers in a variety of line communication applications, it is necessary to consider the various optical fibers currently available.

- The performance characteristics of the various fiber types vary considerably depending upon the materials used in the fabrication process and the preparation technique involved.

- In particular, high performance silica-based fibers for operation in three major wavelength regions 0.8 to 0.9, 1.3 and 1.55 µm are now widely commercially available.

- Complex refractive index profile single-mode fibers including dispersion modified fibers and polarization maintaining fibers are also commercially available and in the former case are starting to find system application within communications.

- Another relatively new area of commercial fiber development is concerned with mid-infrared wavelength range 2 to 5 µm, often employed heavy metal fluoride glass technology.

- However, the fiber products that exists for this wavelength region tend to multimode with relatively high losses and hence at present are only appropriate for special applications.

2.2 DEFINITIONS AND CONCEPTS

2.2.1 Reflection

- We are familiar with light that is reflected from a flat, smooth surface like mirror.

- These reflections shown in Fig. 2.1 are the result of an incident ray and the reflected ray.

- The angle of reflection is determined by the angle of incidence.

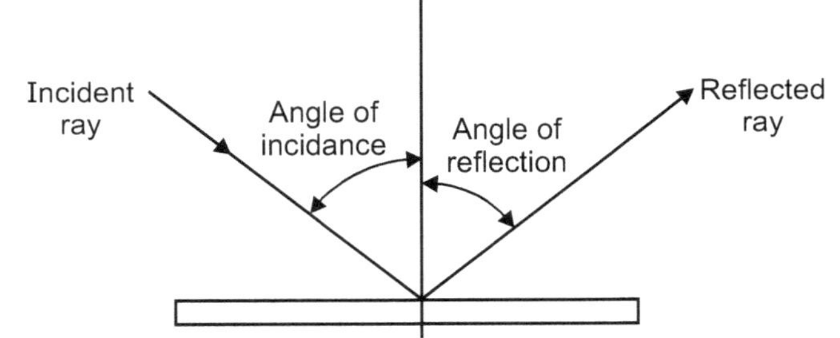

Fig. 2.1: Reflection

Definition :

When the ray incident on a conducting surface it reflects making an same angle with normal is called reflection.

- Reflections in many directions are called **diffuse reflections** and the result of light being reflected by an irregular surface shown in Fig. 2.2.

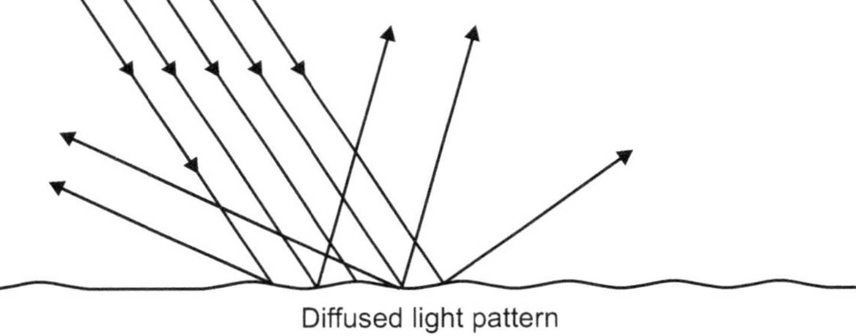

Fig. 2.2: Diffused Reflection

- Example, on a page light which includes all colours is reflected by the rough surface of this page because the roughness is random.

- The reflected light is random i.e. reflects in all directions and because the paper does not absorb much of light, the light seems to be radiate equally from all parts of page.

2.2.2 Refraction

- Another property of light is refraction.

- This is caused by a change in the speed of light as it passes through different mediums such as air, water, glass and the other transparent substances.

Definition :

The deflection of light is called refraction.

- This phenomenon is commonly evident when objects are viewed through glass of water.

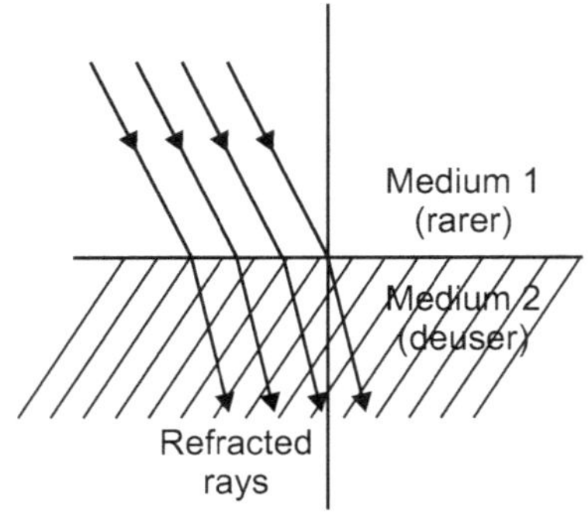

Fig. 2.3: Refraction

- The amount that a light ray is refracted is depends upon the refractive index.
- The **refractive index** can be stated as,

$$n = \frac{c}{v}$$

    where, c = Velocity of light in space

                       v = Velocity of light in specific material

- Each transparent substance has its own refractive index.

2.2.3 Dispersion   `S-16)`

Dispersion is the process of separating light into each of its component frequency.

- It is commonly recognised when sunlight is dispersed into a rainbow of colours by a prism shown in Fig. 2.4.

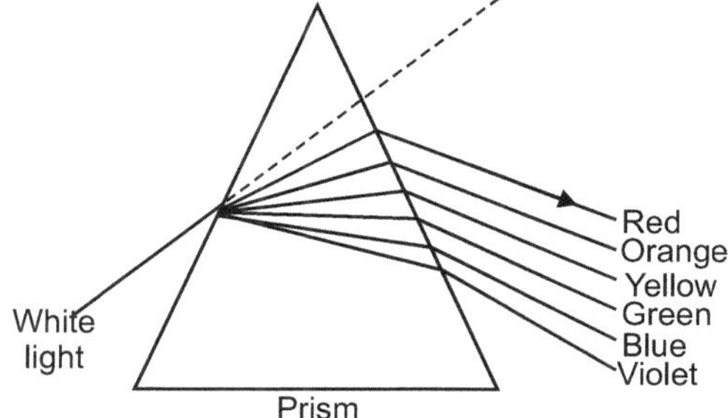

Fig. 2.4: Dispersion

2.2.4 Diffraction

Diffraction is the bending of light as it passes through an opening in an obstacle.

- Diffraction is based on **'Huygens'** principle which states that no matter how small slit is made in an opaque planed, on the side opposite the source would spread out in all directions and no matter how small light source is constructed a sharp shadow cannot be obtained at the edge of the sharp opaque-obstacle.

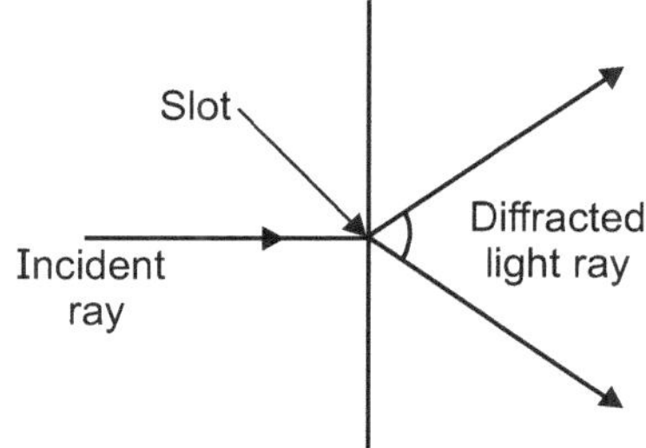

Fig. 2.5: Diffraction

2.2.5 Absorption

Absorption is the process where impurities in the fiber absorb optical energy and dissipate it as a small amount of heat.

- The light becomes dimmer.
- The impurities causing absorption include ions of iron, copper, cobalt and chromium.

2.2.6 Scattering   `(S-16)`

- **Scattering means at a point light spread into all directions.**
- Scattering occurs when light strikes a substance which in turn emits light of its own at the same wavelength as the incident light.

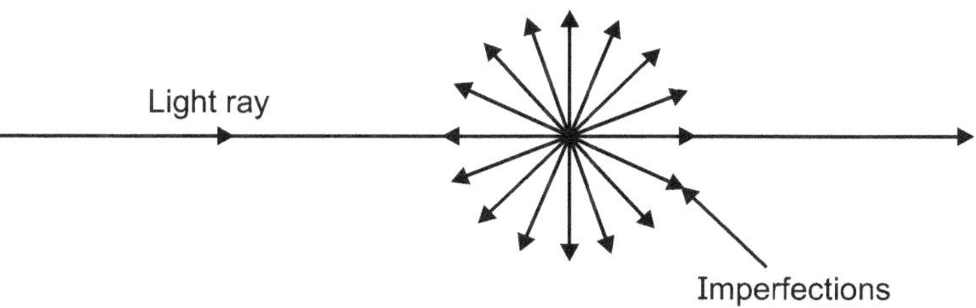

Fig. 2.6: Scattering

- Scattering is the loss of optical energy due to imperfections in the fiber and from the basic structure of the fiber.

2.3 PROPAGATION OF LIGHT (ENERGY) IN FIBER CABLE

Principle :

Light travels at approximately 3×10^8 m/s in free space. If material denser than free space, speed of light is reduced. This reduction in speed as light passes from free space into a denser medium results in refraction (bending) of light.

2.3.1 Snell's Law

It states that incident angle θ_0 is related to exit angle θ_1 by the relation.

$$n_0 \sin \theta_0 = n_1 \sin \theta_1 \qquad \qquad \text{... (2.1)}$$

where, n_0 = Refractive index of air

      n_1 = Refractive index of glass

      θ_0 = Incident angle

      θ_1 = Exit angle

∴ From equation (a)

$$\frac{n_1}{n_2} = \frac{\sin \theta_0}{\sin \theta_1}$$

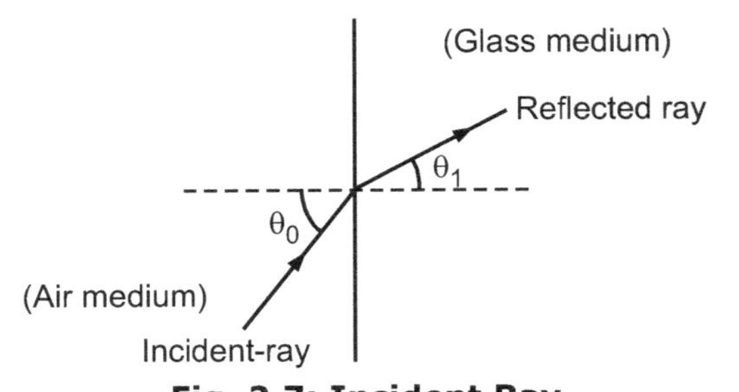

Fig. 2.7: Incident Ray

Now, $n_0 \cong 1$ for air

∴ $n_1 = \dfrac{\sin \theta_0}{\sin \theta_1}$

2.3.2 Critical angle of Fiber Optics

QUESTION

1.   Define w.r.t. optical fiber : critical angle.

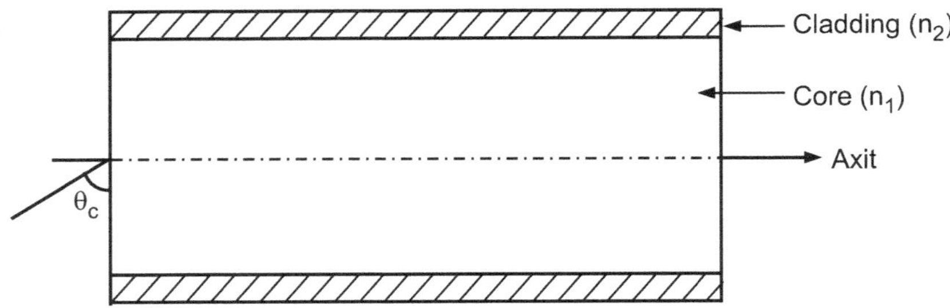

Fig. 2.8: Fiber Optics Cable for Critical Angle

Consider, as shown in Fig. 2.8.

      θ_c = Critical angle

      n_1 = Refractive index of core

      n_2 = Refractive index of cladding $[\because \ n_1 > n_2]$

Definition :

The critical angle of optical fiber is defined as the ratio of refractive index of cladding and core.

$$\sin \theta_c = \frac{n_2}{n_1}$$

$$\boldsymbol{\theta_c = \sin^{-1}\left(\frac{n_2}{n_1}\right)}$$

2.3.3 Acceptance Angle of Optical Fiber

It is the maximum angle to the axis at which light may enter the fiber in order to be propagated.

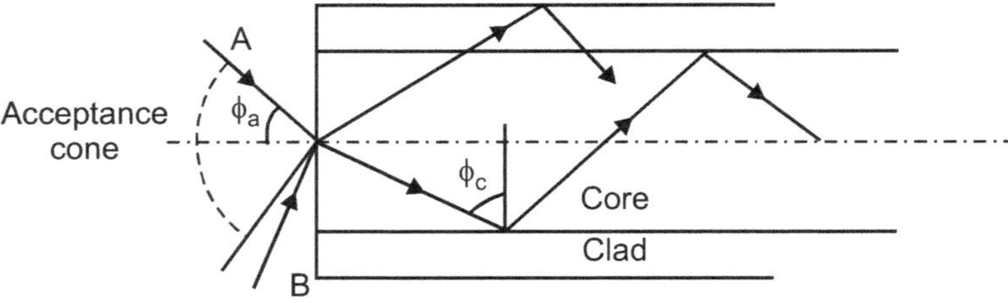

Fig. 2.9

$$\boxed{\theta_A = \sin^{-1} NA}$$

Acceptance cone :

Acceptance cone is defined as the range of angles for rays to be transmitted by total internal reflection within the fiber core.

2.3.4 Aumerical Aperture

QUESTION

1.   Define w.r.t. optical fiber : Numerical aperture.   **(2M)**

- **Numerical Aperture (NA) is the "light gathering ability" of a fiber only light injected into the fiber at angles greater than the critical angle will be propagated.**
- The term NA related to the refractive indices of the core and cladding.

$$\text{NA} = \sqrt{n_1^2 - n_2^2}$$

- NA of a fiber is important because it gives an indication of how the fiber accepts and propagates light.
- A fiber with large NA accepts light well, a fiber with low NA requires highly directional light.

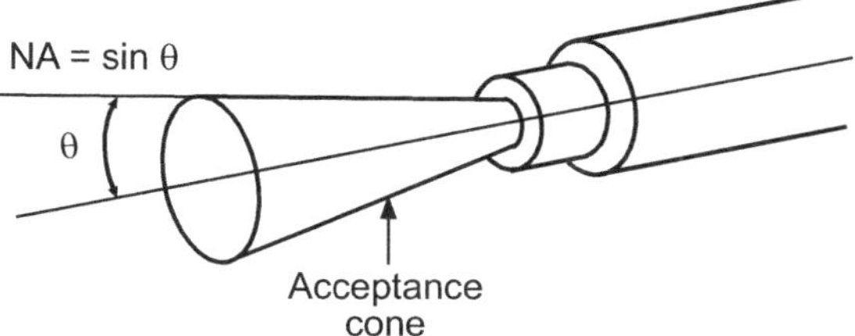

Fig. 2.10: Numerical Aperture

The acceptance angle is related to NA.

$$\theta_{0\,(max)} = \text{arc sin (NA)}$$
$$\text{NA} = \sin \theta_{0\,(max)}$$
$$\therefore \quad \theta_{0\,(max)} = \text{Acceptance angle}$$

NUMERICALS

Example 2.1:

An optical fiber in air has an NA of 0.4. Calculate the acceptance angle.

Solution:

    Given: $\text{NA} = 0.4$

    To find acceptance angle $= ?$

$$\theta_A = \sin^{-1} \text{NA}$$
$$= \sin^{-1} 0.4$$
$$= 23.6°$$

$$\therefore \quad \boxed{\theta_A = 23.6°}$$

Example 2.2:   **(S-15)**

Calculate critical angle of incidence between two substances with different refractive indices $n_1 = 1.4$ and $n_2 = 1.36$.

Solution:

    Given: $n_1 = 1.4$
                        $n_2 = 1.36$

    To find: $\theta_c = ?$

Critical angle:

$$\theta_c = \sin^{-1} \frac{n_2}{n_1}$$
$$= \sin^{-1} \frac{1.36}{1.4}$$
$$= 76.27°$$

$$\boxed{\theta_c = 76.27°}$$

Example 2.3: (S-15)

A silica optical fiber with a core diameter large enough to be considered by ray theory analysis has a core refractive index of 1.50 and a cladding refractive index of 1.47.

Calculate:

(i) Critical angle, (ii) NA of fiber, (iii) Acceptance angle in air for fiber.

Solution:

Given:
$$n_1 = 1.50$$
$$n_2 = 1.47$$

To find:
$$\theta_c = ?$$
$$NA = ?$$
$$\theta_A = ?$$

(i) Critical angle:

$$\theta_c = \sin^{-1} \frac{n_2}{n_1}$$
$$= \sin^{-1} \frac{1.47}{1.50}$$
$$= 78.52°$$

$\therefore$ $\boxed{\theta_c = 78.52°}$

(ii) Numerical aperature:

$$NA = \sqrt{n_1^2 - n_2^2}$$
$$= \sqrt{(1.50)^2 - (1.47)^2}$$
$$= \sqrt{2.25 - 2.16}$$
$$= 0.30$$

$\therefore$ $\boxed{NA = 0.30}$

(iii) Acceptance angle:

$$\theta_A = \sin^{-1} NA$$
$$= \sin^{-1}(0.30)$$
$$= 17.45°$$

$\boxed{\theta_A = 17.45°}$

Example 2.4:

Calculate the NA of a fiber with core index $n_1 = 1.61$ and cladding index $n_2 = 1.55$.

Solution:

Given:
$$n_1 = 1.61$$
$$n_2 = 1.55$$

To find:
$$NA = ?$$

$$NA = \sqrt{n_1^2 - n_2^2}$$
$$= \sqrt{(1.61)^2 - (1.55)^2}$$
$$= \sqrt{2.5921 - 2.4025}$$
$$= 0.435$$

$\therefore$ $\boxed{NA = 0.435}$

Example 2.5: (S-10)

A step index fiber has a numerical aperature of 0.16, a core refractive index of 1.45 and core diameter 90 mm. Calculate:

(i) The acceptance angle, θ_A.

(ii) The refractive index of the cladding.

Solution:

Given:
$$NA = 0.16$$
$$n_1 = 1.45$$

To find:

(i) $\theta_A = ?$

(ii) $n_2 = ?$

(i) Acceptance angle:

$$\theta_A = \sin^{-1} NA$$
$$= \sin^{-1} 0.16$$
$$= 9.21°$$

$\therefore$ $\boxed{\theta_A = 9.21°}$

(ii) Refractive index of cladding:

$$NA = \sqrt{n_1^2 - n_2^2}$$
$$0.16 = \sqrt{(1.45)^2} - \sqrt{n_2^2}$$
$$n_2^2 = (1.45)^2 - (0.16)^2$$
$$n_2 = 1.441$$

$\therefore$ $\boxed{n_2 = 1.441}$

2.4 SPLICING TECHNIQUES   (S-15, 16)

QUESTIONS

1. What are different types of splicing techniques ? Explain fusion splicing techniques. **(4M)**

2. Describe splicing technique. **(4M)**

Splicing is the technique to join the two fiber cables at the bend.

- **A permanent joint formed between two individual optical fibers in the field or factory is known as a fiber splice.**
- Fiber splicing is frequently used to establish longhaul optical fiber links where **smaller fiber lengths to be joined**, so that there is no requirement of repeated connection and disconnection.
- Splices are divided into two categories depending upon the splicing technique utilized.
  (i) Fusion splicing or Welding   **(ii) Mechanical Splicing.**
- **Fusion splicing** is accomplished by applying localised heating at the interface between two butted, prealigned fiber ends causing them to soften and fuse.
- **Mechanical splicing**, in which the fibers are held in alignment by some mechanical means, may be achieved by various methods including the use of tubes around the fiber ends or V-grooves into which the butted fibers are placed (groove splices).
- All these techniques to find to optimize the splice performance through both fiber end preparation and **alignment of the two joint fibers**.
- Note that the insertion losses of fiber splices are generally much less than the possible Frensel reflection loss at a butted fiber-fiber joints.
- This is because there is no large step change in refractive index with the fusion splice as it forms a continuous fiber connection and some method of index matching tend to be utilized with mechanical splices.
- However, especially fusion splicing is at present somewhat difficult process to perform in a field environment and suffers from practical problems.
- Fibers for splicing should have smooth and square end faces.
- In general this end preparation may be achieved using a suitable tool which cleaves the fiber shown in Fig. 2.11.

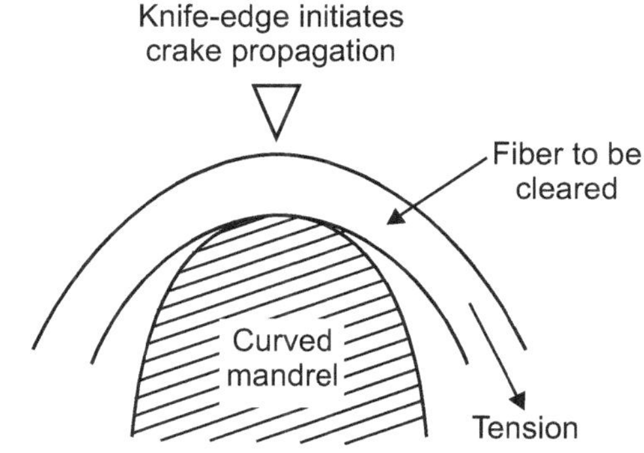

Fig. 2.11: Optical Fiber End Preparation: The Principle of Scribe and Break Cutting

- This process is often referred to as scribe and break or score and break because it includes the scoring of the fiber surface under tension with a cutting tool.
- Fig. 2.11 shows this process with the fiber tensioned around a curved mandrel.
- However, straight pull, scribe and break tools are also utilized, which gives better results.

2.4.1 Fusion Splice   (S-15; W-15)

- The fusion splicing of single fibers, two prepared fiber ends heated to their fusion point with the application of sufficient axial pressure between the two optical fibers.
- Therefore, it is essential that the stripped fiber ends are adequately positioned and aligned in order to achieve good continuity of transmission medium at the junction point.
- Hence, the fibers are usually positioned and clamped with the inspection microscope.
- Heating sources used are microplasma torches and OX hydric microburners. But mostly electric arc is used which gives consistent, easily controlled heat with adaptability for use under field conditions.
- A schematic diagram of basic arc fusion method is given in Fig. 2.12 (a) which shows how the two fibers are welded together.
- Fig. 2.12 (b) shows the development of the basic arc fusion process which involves the rounding of the fiber ends with a low energy discharge before pressing the fibers together and fusing with a stronger arc.

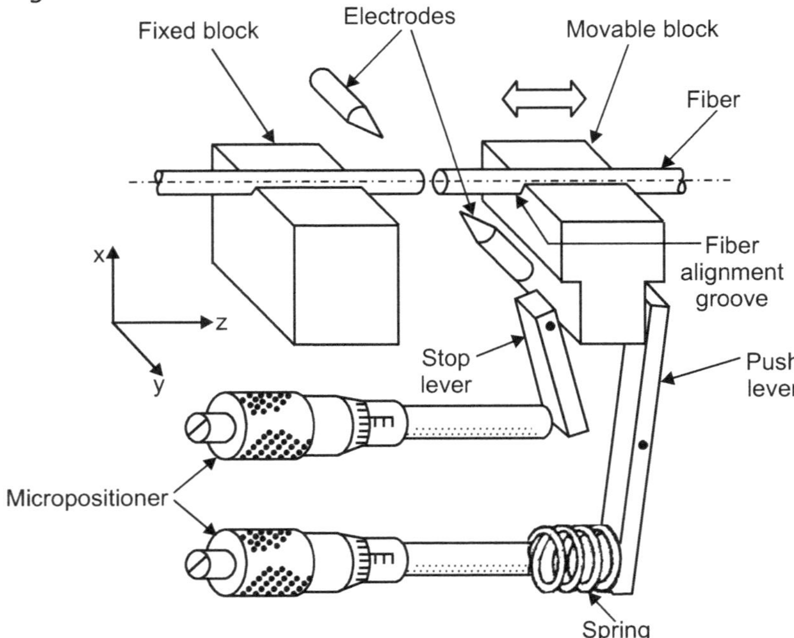

(a) An Example of Fusion Splicing Apparatus

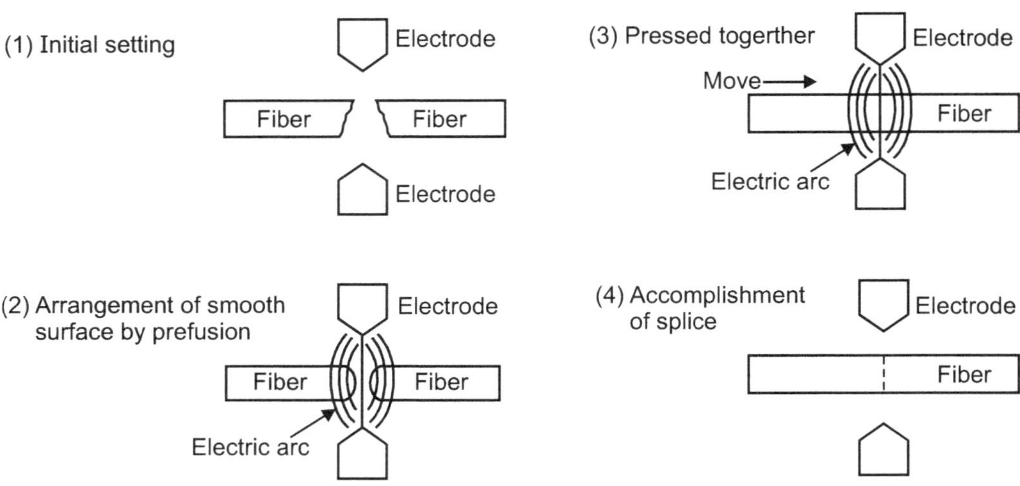

(b) Schematic Illustration of the Prefusion Method for Accurately Splicing Optical Fibers

Fig. 2.12: Electric Arc Fusion Splicing

- This technique known as prefusion, removes the requirement for fiber end preparation.
- Multimode fibers gives average splice losses of 0.09 dB.
- Fusion splicing of single mode fibers with core diameters between 5 and 10 μm presents problems of more critical fiber alignment.
- However, splice inversion losses below 0.3 dB may achieved due to self-alignment phenomenon which partially compensates for any lateral offset.

Self Alignment:

- Self alignment shown in Fig. 2.13 is caused by surface tension effects between the two fiber ends during fusing.
- Mean splice losses of only 0.06 dB have also been obtained with a fully automatic single-mode fiber fusion splicing machine.

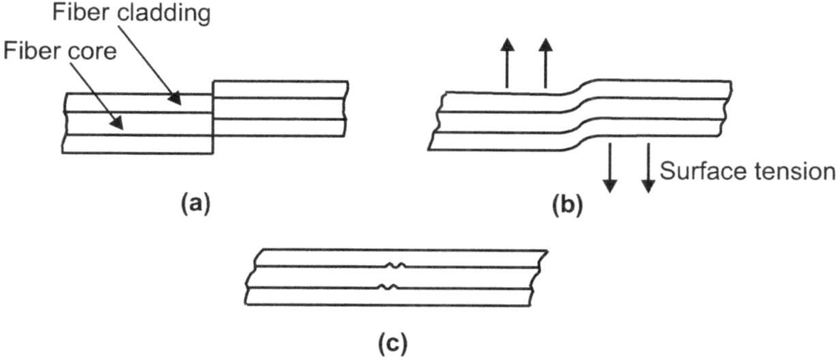

Fig. 2.13: Self-alignment Phenomenon which takes place during Fusion Splicing: (a) Before Fusion, (b) During Fusion; (c) After Fusion

Disadvantages:

1. Heat necessary to fuse the fibers may weaken the fiber in the vicinity of the splice.
2. The tensile strength of the fused fiber may be as low as 30% of that of the uncoated fiber before fusion.

2.4.2 Elastic Tube Splice   (S-15, 16; W-15, 16)

- A common method uses an accurately produced rigid alignment tube into which the prepared fiber ends are permanently bonded.
- This snug tube splice shown in Fig. 2.14 (a) and many utilize a glass or ceramic capillary with inner diameter just large enough to accept the optical fibers.

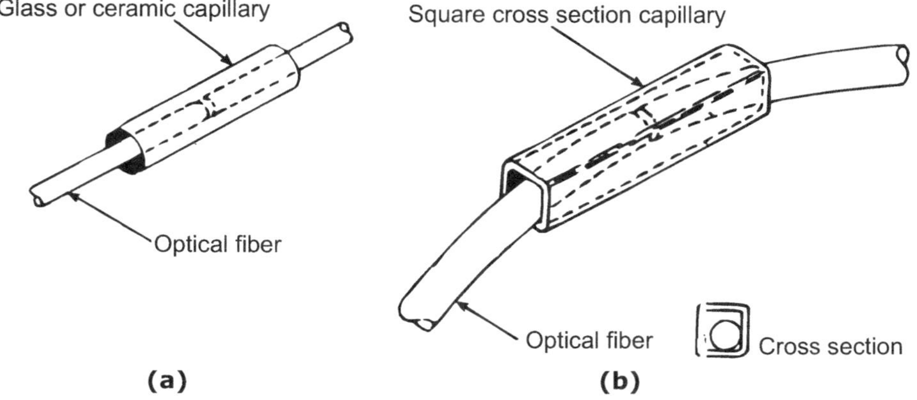

Fig. 2.14: Techniques for Tube Splicing of Optical Fibers: (a) Snug Tube Splice, (b) Loose Tube Splice utilizing Square Cross-section Capillary

- Transparent adhesive is injected through a transverse bore in the capillary to give mechanical sealing and index machining of the splice.
- Average insertion losses as low as 0.1 dB have been obtained with multimode graded index and single mode fibers using ceramic capillaries.
- However, snug tube splices exhibit problems with capillary tolerance requirement.
- Hence, as a commercial products they may exhibit losses of upto 0.5 dB.
- A elastic tube splicing technique which avoids the critical tolerance requirements of the snug tube splice shown in Fig. 2.14 (b).

- This loose tube splice uses an oversized square section metal tube which easily accepts the prepared fiber ends.
- Transparent adhesive is first inserted into the tube followed by the fiber.
- The splice is self-aligning when the fiber are curved in the same plane, forcing the fiber ends simultaneously into the same corner of the tube shown in Fig. 2.14 (b).

2.4.3 V-groove Splices   (S-15, 16; W-15)

- Other common mechanical splicing techniques involves the use of grooves to secure the fibers to be jointed.
- A simple method utilizes a V-groove into which the two prepared fiber ends are pressed.
- The V-groove splice shown in Fig. 2.15 (a) gives alignment of the prepared fiber ends through insertion in the groove.
- The splice is made permanent by securing the fibers in the V-groove with epoxy resin.

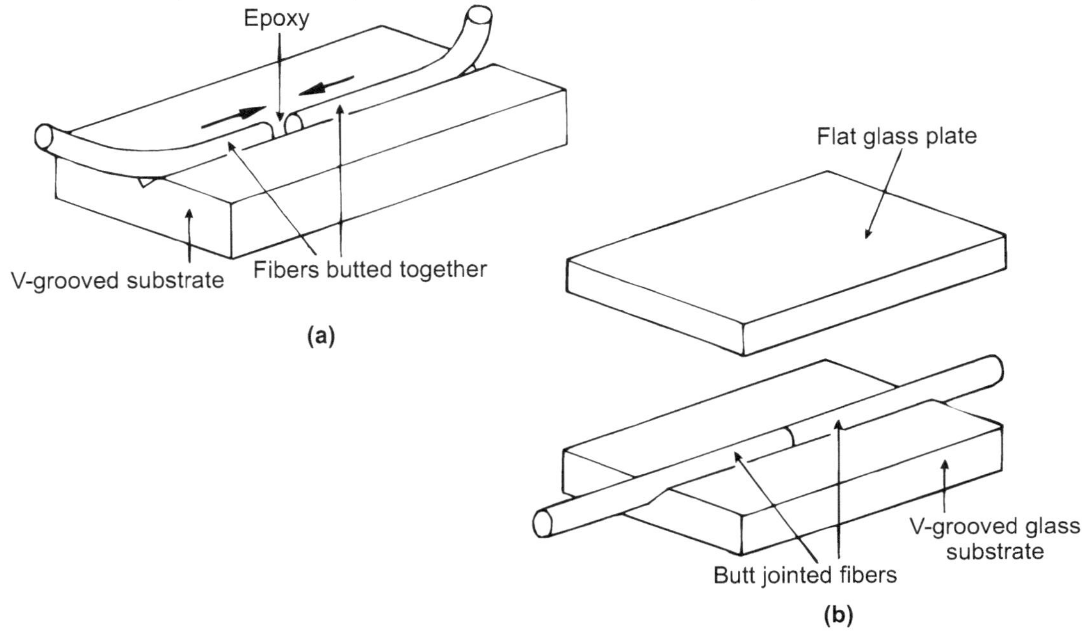

Fig. 2.15: V-groove Splices

- V-groove formed by sandwitching the butted ends between a V-groove glass substrate and a flat glass retainer plane shown in Fig. 2.15 (b).

2.5 FIBER CONNECTOR   (S-16)

- Fiber connectors are special mechanical assemblies that allow fiber optic cable to be connected to one another.
- A variety of connectors provide a convenient way to splice cables and attach them to transmitters, receivers and repeaters.

Properties:

1. Connectors force better alignment of the cables.
2. Two ends of cables must be aligned with precision so that excessive light is not lost. Otherwise, a splice or connection will introduce excessive attenuation.
3. Fiber optic connectors are the optical equivalent of electrical plugs and sockets.
4. They are mechanical assemblies that hold the ends of the cable and cause them to be accurately aligned with another cable.
5. Most of them either snap together or have threads which allow the two pieces to be screwed together.
6. Optical connectors corrects the problems of misalignment.

2.5.1 Ferrule Connector

- A typical Ferrule connector is shown in Fig. 2.16 (a).
- One end of the connector called Ferrule holds the fiber securely in place.

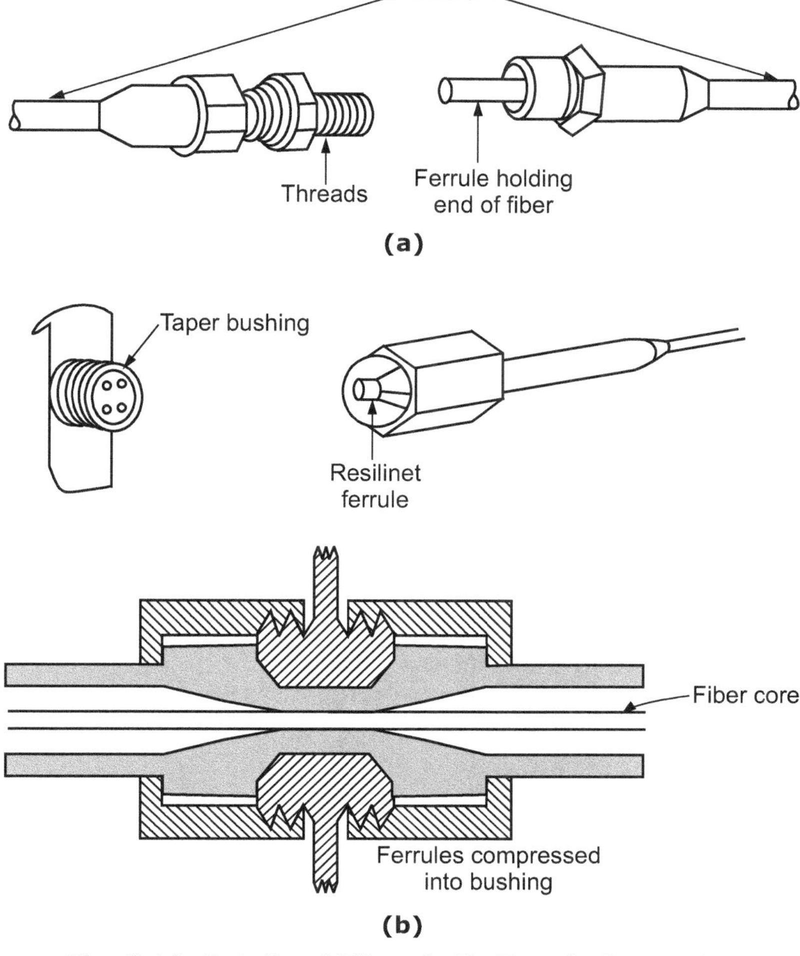

Fig. 2.16: Details of Fiber-Optic Ferrule Connector

- A matching fitting holds the other fiber securely in place.
- When the two are screwed together, the ends of the fiber touch, providing a low-loss coupling.
- Fig. 2.16 (b) shows how the connector aligns the fibers.
- Connectors are normally used at the end of the cable applied to the light source or at the end of the cable connected to the photo detectors.
- Connectors are also used at the repeater units where the light is picked-up, converted into an electrical pulse, amplified and reshaped and then used to create a new pulse to continue the transmission over a long line.

2.6 LOSSES IN OPTICAL FIBER

- Here, we shall continue the discussion of optical fibers by answering two very important questions:
  1. What are the loss or signal attenuation mechanism in a fiber ?
  2. Why and to what degree do optical signals get distorted as they propagate along a fiber ?
- Signal attenuation known as fiber loss or signal loss is one of the most important properties of fiber optic, because it determines largely the maximum repeaterless separation between a transmitter and a receiver.
- Since the repeaters are expensive to fabricate, install and maintain, the degree of attenuation in a fiber has a large influence on system cost.
- Also equal importance is signal distortion.
- The distortion mechanisms in a fiber cause optical signal pulses to broaden as they travel along a fiber.
- If these pulses travel sufficiently long distance, they will eventually overlap with neighouring pulses, thereby creating errors in the receiver output.

- The signal distortion mechanisms thus limits the information carrying capacity of a fiber.
- The predominant losses in optical fiber cable are:
  1. Absorption loss   2. Scattering loss   3. Dispersion loss
  4. Radiation loss   5. Coupling loss

Attenuation

QUESTION

1. Describe attenuation in optical fiber. 

- The main specification of a fiber optic cable is its attenuation.
- **Attenuation means the loss of light energy as the light pulse travels from one end of cable to the other.**
- The light pulse of specific amplitude is applied to one end of the cable, but the light pulse output at the other end will be much lower in amplitude.
- The intensity of light at the output is low because of various losses in the cable.
- The reasons for the losses in light intensity over the length of the cable is due to light.
  - Absorption,
  - Scattering, and
  - Dispersion.
- The amount of attenuation varies with the type of cable and its size.
- Glass has less attenuation than plastic.
- Wider cores have less attenuation than narrow cores.
- **Attenuation is directly proportional to the length of the cable.**
- It is obvious that the longer the distance the light has to travel, the greater the loss due to absorption, scattering and dispersion.
- Doubling the length of cable doubles the attenuation and so on.
- The attenuation of a fiber optic cable is expressed in logarithmic unit of decibel.

$$\therefore \qquad dB = 10 \log_{10} \frac{P_o}{P_{in}} \qquad \ldots (2.2)$$

where, P_o = Power output

                 P_i = Power in

The decibel is used for comparing two power levels, may be defined for a particular optical wavelength as the ratio of the input (transmitted) optical power P_i into a fiber to the output (received) optical power P_o from the fiber.

Table 2.1 shows the percentage of output power for various decibel losses.

For example, 3 dB represents half power.

In other words, 3 dB loss means only 50% of the input appears at the output, the other 50% of the power is lost in the cable.

Table 2.1: Decibel Loss Table

Loss (dB)	Power Output (%)
1	79
2	63
3	50
4	40
5	31
6	25
7	20
8	14
9	12
10	10
20	1
30	0.1
40	0.01
50	0.001

- The higher the decibel figure, the greater the attenuation and loss.
- A 30 dB loss means only one-thousandth of the input power appears at the end.
- In fiber optic communications, the attenuation is usually expressed in decibels per unit length i.e. dB/km.

$$\therefore \qquad \alpha_{dB}\, L \;=\; 10\,\log_{10} \frac{P_i}{P_o} \qquad\qquad \text{... (2.3)}$$

where,

$$\alpha_{dB} = \text{Signal attenuation per unit length in decibels}$$
$$L = \text{Fiber length}$$

Example 2.6:

When the mean optical power launched into an 8 km length of the fiber is 120 μW, the mean optical power at the fiber output is 3 μW.

Determine:

(a) The overall signal attenuation or loss in decibels.

(b) The signal attenuation per km for the fiber.

(c) The overall signal attenuation for a 10 km optical link using the same fiber with splices at 1 km intervals, each giving an attenuation of 1 dB.

(d) The numerical input/output power ratio in (C).

Solution:

(a) Use equation 2.1, the overall attenuation in decibels through the fiber is,

$$\text{Signal attenuation} = 10\,\log_{10} \frac{P_i}{P_o}$$
$$= 10\,\log_{10} \frac{120 \times 10^{-6}}{3 \times 10^{-6}}$$
$$= \log_{10} 40$$
$$= \mathbf{16\ dB}$$

(b) The signal attenuation per km for the fiber may be obtained by dividing the results in (a) by the fiber length. Use equation 6.2.

$$\alpha_{dB}\, L = 16\ dB$$
$$\therefore \qquad \alpha_{dB} = \frac{16}{8}$$
$$= \mathbf{2\ dB/km}$$

(c) α_{dB} = 2 dB/km, the loss incurred along 10 km of fiber is,

$$\alpha_{dB}\, L = 2 \times 10 = \mathbf{20\ dB}$$

However, the link has nine splices (at 1 km intervals) each with an attenuation of 1 dB. Thus, loss due to splices is 9 dB. Hence, overall signal attenuation for link is,

$$\text{Signal attenuation} = 20 + 9$$
$$= 29\ dB$$

(d) To obtain a numerical value for input/output power ratio may be,

$$\frac{P_i}{P_o} = 10^{(dB/10)}$$
$$= 10^{29/10}$$
$$= \mathbf{794.3}$$

2.7 ABSORPTION LOSS (S-16)

- **Material absorption is a loss mechanism related to the material composition and the fabrication process for the fiber, which results in dissipation of some of the transmitted optical power as heat in the waveguide.**
- The absorption of the light may be:

(i) Intrinsic:

Caused by the interaction with one or more of the components of glass.

(ii) Extrinsic:

Caused by impurities within the glass.

2.7.1 Intrinsic Absorption

- A pure silicate glass has small intrinsic absorption due to its basic material structure in the near infra-red region.
- However, it has two major intrinsic absorption mechanisms at optical wavelengths which leave a low intrinsic absorption window over the 0.8 to 1.7 μm wavelength range shown in Fig. 2.17.

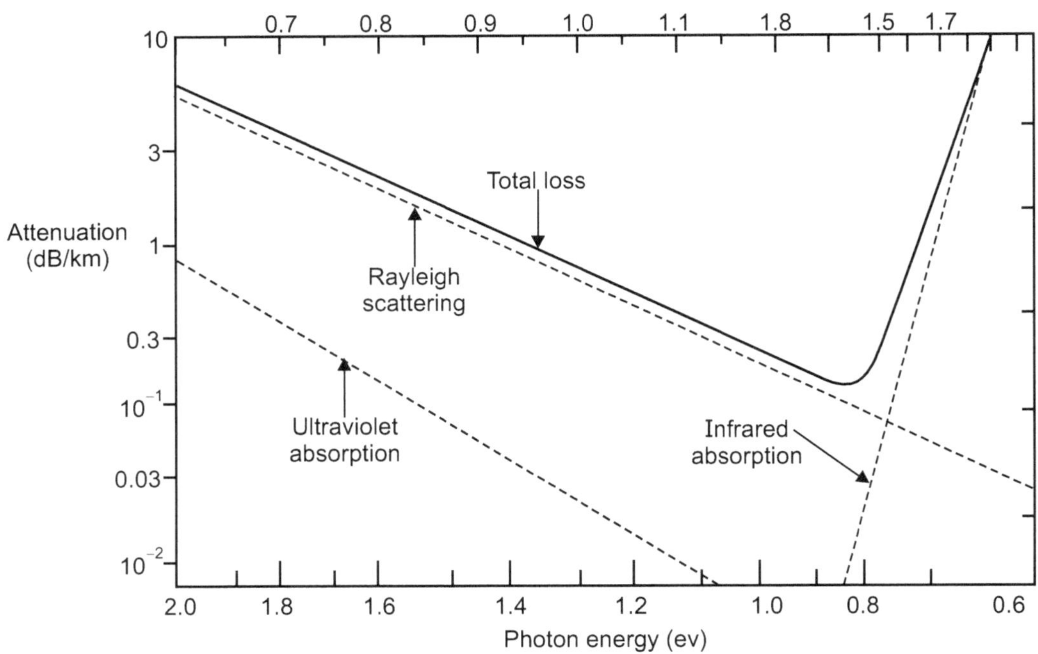

Fig. 2.17: The Attenuation Spectra for the Intrinsic Loss Mechanism in pure GeO$_2$ – SiO$_2$ Glass

- Fig. 2.17 shows a possible optical attenuation against wavelength characteristics for absolutely pure glass.
- It shows that there is a fundamental absorption edge, the peaks of which are centred in the ultraviolet wavelength region.
- Also in the infrared and far-infrared, at wavelength above 7 μm fundamentals of absorption bands from the interaction of photons with molecular variations within the glass occur.
- This give absorption peaks which again extends into the window region.
- The strong absorption band occurs due to oscillations of structural units such as Si-O (9.2 μm), P-O (8.1 μm), B-O (7.2 μm) and Ge-O (11 μm) within the glass.
- Hence, above 1.5 μm the tails of these largely for infrared absorption peaks tend to cause most of the pure glass losses.
- However, effects of both these processes may be minimized by suitable choice of both core and cladding compositions.
- In some non-oxide glasses as fluorides and chlorides, the infrared absorption peaks occur at much longer wavelengths which are well into the far-infrared (upto 50 μm), gives less attenuation to longer wavelength transmission compared with oxide glass.

2.7.2 Extrinsic Absorption

- **Extrinsic absorption result from the presence of impurities.**
- Some common metallic impurities found in glasses shown in Table 2.2 together with the absorption losses caused by one part in 10^9.

**Table 2.2: Absorption losses caused by some
of the more common metallic ion impurities in glasses**

Ion Impurities	Peak Wavelength (nm)	One Part in 10^9 (dB/km)
C_r^{3+}	625	1.6
C_+^{2}	685	0.1
C_u^{2+}	850	1.1
Fe^{2+}	1100	0.68
Fe^{3+}	400	0.15
Ni^{2+}	650	0.1
Mn^{2+}	460	0.2
V^{4+}	725	2.7

- From Table 2.2, it may be noted that certain of these impurities, namely chromium and copper, in their worst valence state can cause attenuation in excess of 1 dB/km in the near-infrared region.

- Transition element contamination may be reduced to acceptable levels (i.e. one part in 10^{10}) by glass refining techniques such as vapour-pulse oxidation which eliminates the effects of these metallic impurities.

- However, another major **extrinsic loss mechanism** is caused by absorption due to water (i.e. OH ion) dissolved in the glass.

- These hydroxyl groups are bonded into the glass structure and have fundamental stretching vibrations which occur at wavelengths between 2.7 and 4.2 µm depending one group position in the glass network.

- The fundamental vibrations give rise to over tones appearing harmonically at 1.38, 0.95 and 0.75 µm shown in Fig. 2.18.

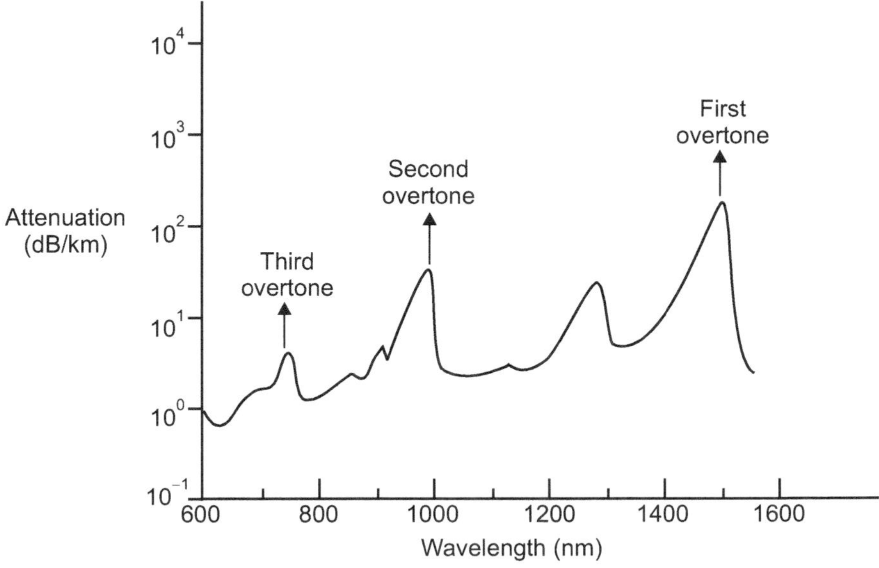

Fig. 2.18: The Absorption Spectrum for the Hydroxyl (OH) Group in Silica

- Fig. 2.18 shows the absorption spectrum for the hydroxyl group in silica.

- The combinations between the overtones and the fundamental SiO_2 vibration occur at 1.24, 1.13 and 0.88 µm completing the spectrum.

- It may also be observed that, the only significant absorption band in the region below a wavelength of 1 μm is the second overtone at 0.95 μm which causes attenuation at about 1 dB/km for one part per million of hydroxyl.
- At longer wavelengths the first overtone at 1.38 μm and its sideband at 1.24 μm are strong absorbers gives attenuation of 2 dB/km ppm and 4 dB/km ppm respectively.

2.8 SCATTERING LOSS (W-15; S-16)

> **QUESTION**
>
> 1. Explain the losses due to scattering in FOC.

- **Scattering loss is one type of attenuation.**
- Scattering is the loss of optical energy due to imperfections in the fiber and from the basic structure of the fiber. It scatters the light in all directions shown in Fig. 2.19.

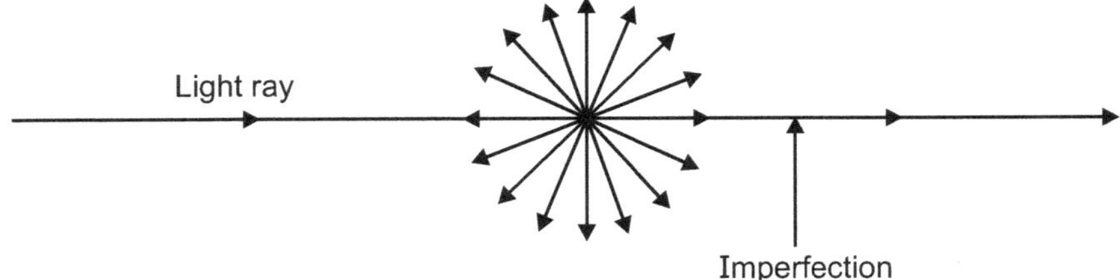

Fig. 2.19: Scattering of Light

Scattering is inversely proportional to the forth power of the wavelength $\left(\frac{1}{\lambda^4}\right)$.

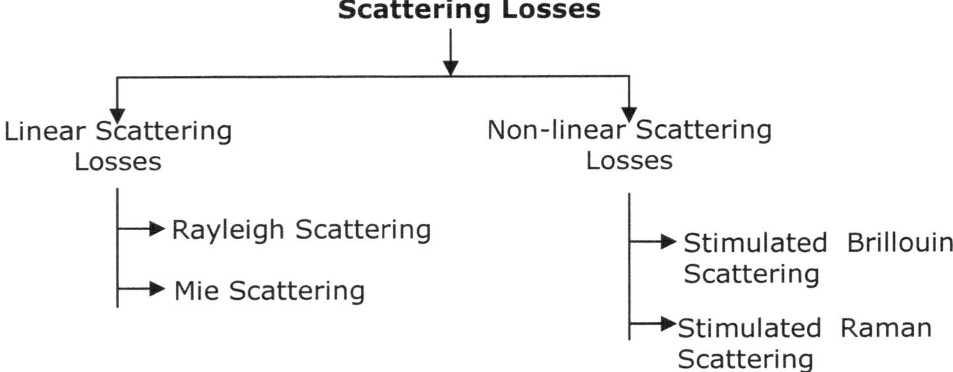

2.9 LINEAR SCATTERING LOSS

- Linear scattering mechanisms cause the transfer of some or all of the optical power contained within one propagating mode to be transferred linearly into different mode.
- This process tends to result in attenuation of the transmitted light as the transfer may be a leaky or radiation mode which does not continue to propagate within the fiber core, but is radiated from the fiber.
- Note that with all linear processes there is no change of frequency on scattering.
- Linear scattering categorized in two types:
  - **(i) Rayleigh scattering.**
  - **(ii) Mie scattering.**
- Both results from the non-ideal physical properties of the manufactured fiber which are difficult and in certain cases impossible to eradicate at present.

2.9.1 Rayleigh Scattering

- **Rayleigh scattering is the dominant intrinsic loss mechanism in the low absorption window between the ultraviolet and infrared absorption tails.**
- It results from inhomogeneities of a radom nature occurring on a small scale compared with the wavelength of light.

- These homogeneities manifest themselves as refractive index fluctuations, arise from density and compositional variations which are frozen into the glass lattice on cooling.
- These compositional variations may be reduced by improved fabrications, but the index fluctuations caused by the freezing in of density inhomogeneities are fundamental and cannot be avoided.
- **The subsequent scattering due to the density fluctuations, which is almost in all directions, produces an attenuation proportional to $\frac{1}{\lambda^4}$ following the Rayleigh scattering formula.**
- For a single component glass this is given as,

$$\gamma_R \ = \ \frac{8\pi^3}{3\lambda^4} \, n^8 \rho^2 B_c \, kT_f \qquad \qquad \ldots (2.4)$$

where,

γ_R = Rayleigh scattering coefficient.

λ = Optical wavelength.

n = Refractive index of the medium.

ρ = Average photoelastic coefficient.

B_c = Isothermal compressibility at fictive temperature T_f.

k = Boltzmann's constant.

Here, **the fictive temperature is defined as the temperature at which the glass can reach a state of thermal equilibrium and is closely related to the anneal temperature.**

- Also, the Rayleight scattering coefficient is related to the transmission loss factor (i.e. transmissivity) of the fiber $\mathscr{L}$ following the reaction

$$\mathscr{L} \ = \ e \times \rho \, (-\gamma_R L) \qquad \qquad \ldots (2.5)$$

where,

L = Length of the fiber

From equation (2.5), it is apparent that the fundamental component of Rayleigh scattering is strongly reduced by operating at the longest possible wavelength.

- The theoretical attenuation due to Rayleigh scattering in silica at wavelengths of 0.63, 1.00 and 1.30 μm respectively, these theoretical results are reasonable agreement with experimental work.
- The predictable attenuation due to Rayleigh scattering against wavelength is indicated by broken line on the attenuation characteristics shown in Fig. 2.20

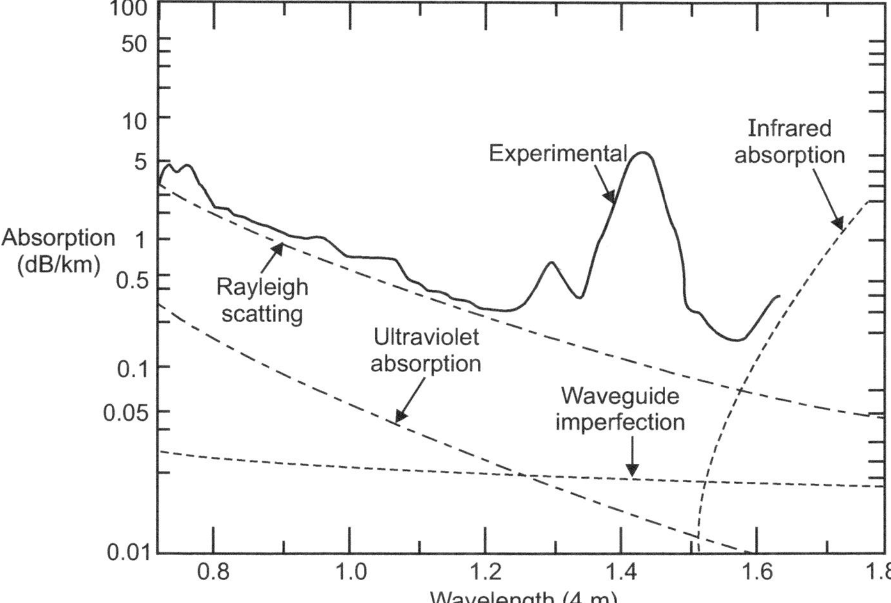

Fig. 2.20: Attenuation Spectrum for an Ultra-low-loss Single Mode Fiber with the Calculated Attenuation Spectra for some of the Loss Mechanism contributing to overall Fiber Attenuation

Example 2.7:

Silica has an estimated fictive temperature of 1400 K with an isothermal compressibility of 7×10^{-11} m^2/N. The refractive index and the photoleastic coefficient for silica and 1.46 and 0.286 respectively.

Determine the theoretical attenuation in decibels per km due to the fundamental Rayleigh scattering in silica at optical wavelengths of 0.63, 1.00 and 1.30 μm. Boltzman's constant is 1.381×10^{-2} J/K.

Solution:

Given:

$$T_f = 1400 \text{ K}$$
$$\beta_c = 7 \times 10^{-11} \text{ m}^2/\text{N}$$
$$n = 1.46$$
$$\rho = 0.286$$
$$\lambda = 0.63, 1 \text{ and } 1.30 \text{ μm}$$
$$k = 1.381 \times 10^{-23} \text{ J/K}$$

The Rayleigh scattering coefficient obtained from equation (6.3),

$$\gamma_R = \frac{8\pi^3 n^8 \rho^2 \beta_c \, k \, T_f}{3\lambda^4}$$

$$\therefore \quad \gamma_R = \frac{8\pi^3 \times (1.46)^8 \times (0.286)^2 \times 7 \times 10^{-11} \times 1.381 \times 10^{-3} \times 1400}{3 \times \lambda^4}$$

$$= \frac{248.15 \times 20.65 \times 0.082 \times 7 \times 10^{-11} \times 1.381 \times 10^{-23} \times 1400}{3\lambda^4}$$

$$= \frac{1.895 \times 10^{-28}}{\lambda^4} \text{ / m}$$

At a wavelength 0.63 μm

$$\gamma_R = \frac{1.895 \times 10^{-28}}{(0.63 \times 10^{-6})^4}$$

$$= \frac{1.895 \times 10^{-28}}{0.158 \times 10^{-24}}$$

$$= \mathbf{1.199 \times 10^{-3} \text{ /m}}$$

The transmission loss factor one km of fiber obtained from equation (6.4).

$$\mathscr{L}_{km} = e \times \rho \, (-\gamma_R L)$$

$$= e \times \rho \, (-1.199 \times 10^{-3} \times 10^3)$$

$$= \mathbf{0.301}$$

The attenuation due to Rayleigh scattering in dB/m obtained from equation (6.1).

$$\text{Attenuation} = 10 \log_{10} \left(\frac{1}{\mathscr{L}_{km}}\right)$$

$$= 10 \log_{10} 3.322$$

$$= \mathbf{5.2 \text{ dB/km}}$$

At a wavelength of 1.00 μm

$$\gamma_R = \frac{1.895 \times 10^{-28}}{10^{-24}}$$

$$= \mathbf{1.895 \times 10^{-4} \text{ /m}}$$

Using equation (6.4),

$$\mathscr{L}_{km} = e \times \rho \, (-1.895 \times 10^{-4} \times 10^3)$$

$$= e \times \rho \, (-0.1895)$$

$$= \mathbf{0.827}$$

From equation (6.1),

$$\text{Attenuation} = 10 \log_{10} 1.209$$
$$= 0.8 \text{ dB/km}$$

At a wavelength of 1.30 µm

$$\gamma_R = \frac{1.895 \times 10^{-28}}{2.856 \times 10^{-24}}$$
$$= \mathbf{0.664 \times 10^{-4}}$$

Using equation (6.4),

$$\mathscr{L}_{km} = e \times \rho \left(-0.664 \times 10^{-4} \times 10^{3}\right)$$
$$= \mathbf{0.936}$$

From equation (6.1),

$$\text{Attenuation} = 10 \log_{10} 1.069$$
$$= \mathbf{0.3 \text{ dB/km}}$$

2.9.2 Mie Scattering

- Linear scattering also occur at inhomogeneities which are compatible in size to the guided wavelength.

- These results from the non-perfect cylindrical structure of the waveguide and may be caused by fiber imperfections such as irregularities in the core-cladding interface, core-cladding refractive index differences along with fiber length, diameter fluctuations, strains and bubbles.

- When the scattering inhomogeneity size is greater than $\lambda/10$, the scattered intensity which has an angular dependence can be very large.

- **The scattering created by such inhomogeneities is mainly in forward direction, is called Mie scattering.**

- Depending on the fiber material, design and manufacture, Mie scattering can cause losses.

- The inhomogenities can be reduced by:
  (i)   Removing imperfections due to the glass manufacturing process.
  (ii) Carefully controlled extrusion and coating of the fiber.
  (iii) Increasing the fiber guidance by increasing the relative refractive index difference.
  So that it is possible to reduce Mie scattering to insignificant levels.

2.10 NON-LINEAR SCATTERING

- Optical wavelength do not always behave as completely linear channels whose increase in output optical power is directly proportional to the input optical power.

- Usually at high optical power levels, several non-linear effects occur, which is in the case of scattering cause disproportionate attenuation.

- This non-linear scattering causes the optical power from one mode to the transferred in either the forward or backward direction to the same or other modes at different frequency.

- It depends critically on the optical power density within the fiber and hence become significant only above threshold power levels.

- There are **two important types of non-linear scattering** within optical fibers are:
  (i)   Stimulated Brillouin and
  (ii) Raman scattering

- Both of these types usually observed only at high optical power densities in long single mode fibers.

- These scattering mechanisms give optical gain but with a shift in frequency, thus contributing to attenuation for light transmission at specific wavelength.

- Note that such non-linear phenomena can also be used to give optical amplification in the context of integrated optical techniques.

2.10.1 Stimulated Brillouin Scattering

- **Stimulated Brillouin Scattering (SBS) regarded as the modulation of light through thermal molecular vibrations within the fiber.**
- The scattered light appears as upper and lower sidebands which are separated from the incident light by the modulation frequency.
- The incident photon in this scattering process produces a photon of acoustic frequency as well as a scattered photon.
- This produces an optical frequency shift which varies with the scattering angle because the frequency of sound wave varies with acoustic wavelength.
- Frequency shift is a maximum in the backward direction reducing to zero in the forward direction making SBS and backward process.
- Brillouin scattering is significant only above a threshold power density, assuming that the polarization state of the transmitted light is not maintained i.e. the threshold power P_B is given by,

$$P_B = 4.4 \times 10^{-3} \, d^2 \lambda^2 \alpha_{dB} \, \gamma \text{ Watts} \qquad \qquad \dots (2.6)$$

where,

$$d = \text{Fiber core diameter in } \mu m.$$
$$\lambda = \text{Operating wavelength in } \mu m.$$
$$\alpha_{dB} = \text{Fiber attenuation in dB/km.}$$
$$\gamma = \text{Source bandwidth in GHz.}$$

Equation (6.11) allows the determination of the threshold optical power which must be launched into a single-mode optical fiber before SBS occurs.

2.10.2 Stimulated Raman Scattering

- **Stimulated Raman Scattering (SRS)** is similar to SBS except that a high frequency optical photon rather than an acoustic photon is generated in the scattering process.
- SRS can occur in both forward and backward directions in the optical fiber and may have an optical power threshold of upto three orders of magnitude higher than the Brillouin threshold in a particular fiber.
- Using some criteria as those specified in SBS threshold given in equation (2.6), it may be shown that the threshold optical power for SRS in a long single mode fiber is given by,

$$P_R = 5.9 \times 10^{-2} \, d^2 \lambda \, \alpha_{dB} \text{ Watts} \qquad \qquad \dots (2.7)$$

- SBS and SRS are not usually observed in multimode fibers because of their large core diameter make the threshold optical power levels extremely high.
- Note that the threshold optical powers for both these scattering mechanisms may be increased by suitable adjustment of the other parameters in equations (2.6) and (2.7).
- Operation at the longest possible wavelength is advantageous.

Example 2.8:

A long single-mode optical fiber has an attenuation of 0.5 dB/km when operating at a wavelength of 1.3 μm. The fiber core diameter is 6 μm and the laser source bandwidth is 600 MHz. Compare the threshold optical powers for Stimulated Brillouin and Raman Scattering within the fibers at the specified wavelength.

Solution:

Given:

$$d = 6 \, \mu m$$
$$\lambda = 1.3 \, \mu m$$
$$\alpha_{dB} = 0.5 \text{ dB/km}$$
$$\gamma = 600 \text{ MHz}$$

The threshold optical power for SBS is,

$$P_B = 4.4 \times 10^{-3} \; d^2 \lambda^2 \; \alpha_{dB} \; \nu$$
$$= 4.4 \times 10^{-3} \times 6^3 \times 1.3^2 \times 0.5 \times 0.6$$
$$= \textbf{80.3 mV}$$

The threshold optical power for SRS is,

$$P_R = 5.9 \times 10^2 \; d^2 \; \lambda \; \alpha_{dB}$$
$$= 5.9 \times 10^{-2} \times 6^2 \times 1.3 \times 0.5$$
$$= \textbf{1.38 W}$$

2.11 DISPERSION LOSS 

QUESTIONS	
1. Explain the intramodal and intermodal dispersion.	**(4M)**
2. What is dispersion ? Describe dispersion.	**(4M)**

Dispersion of light is the distortion of light pulse and it is broadening of light pulse and merging of pulse into another pulse.

- Dispersion of the transmitted optical signal causes distortion for both digital and analog transmission along optical fibers.

- **Digital modulation used for optical fiber, then dispersion mechanisms within the fiber cause broadening of the transmitted light pulses as they travel along the channel.**

- From Fig. 2.21, it may be observed that each pulse broadens and overlaps with its neighbours, becoming indistinguishable at the receiver input. The effect is known as **Intersymbol Interference (ISI).**

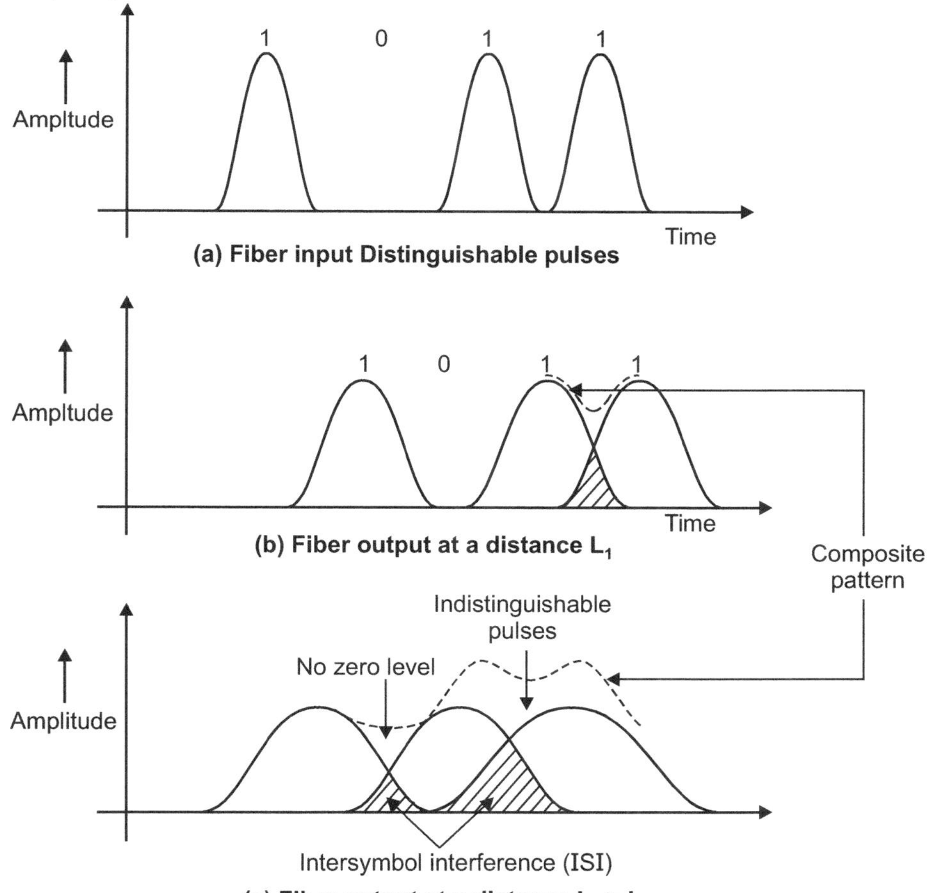

(a) Fiber input Distinguishable pulses

(b) Fiber output at a distance L_1

(c) Fiber output at a distance $L_2 > L_1$

Fig. 2.21: The Digital Bit Pattern 1011 of the Broadening of Light Pulses as they are transmitted along a Fiber

- Thus, number of errors may be encountered on the digital optical channel as the ISI becomes more pronounced.
- The error rate is also a function of the signal attenuation on the link and the subsequent signal to noise ratio (SNR) at the receiver.
- Signal dispersion alone limits the maximum possible bandwidth attainable with a particular optical fiber to the point where individual symbols can no longer be distinguished.
- For no overlapping of light pulses down on an optical fiber link the digital bit rate B_T must be less than the reciprocal of the broadened pulse duration (2τ). Hence,

$$B_T \leq \frac{1}{2\tau}$$

- This assumes that the pulse broadening due to the dispersion on the channel is τ, which dictates the input pulse duration that is also τ.
- Another more accurate estimate of the maximum bit rate for an optical channel with dispersion may be obtained by considering the light pulses at the output to have Gaussian shape with an r.m.s. width 6.
- **The effect of dispersion affects the maximum bit rate and hence the bandwidth.**
- A longer fibers results in more pulse broadening and hence leads to lower bandwidth.
- The dispersion also depends on the wavelength and bandwidth.
- For higher wavelength dispersion is less and hence bandwidth is more.
- **Types of Dispersion:**

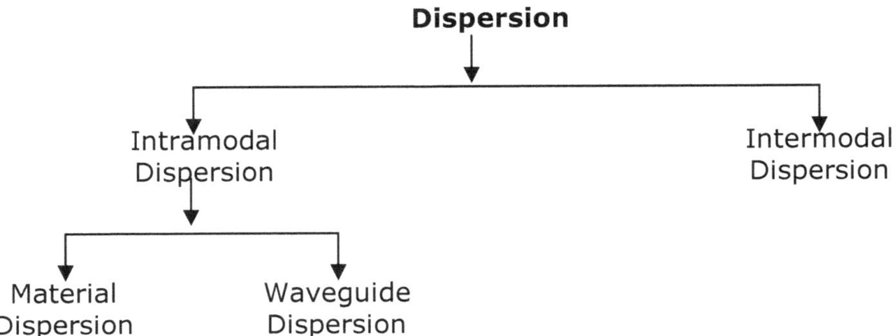

2.11.1 Intramodal Dispersion   (W-15, 16)

- Intramodal or chromatic dispersion may occur in all types of optical fiber and results from the finite spectral linewidth of the optical source.
- Since optical sources do not emit just a single frequency but a band of frequencies then there may be propagation delay differences between the different spectral components of the transmitted signal.
- This causes broadening of each transmitted mode and hence intramodal dispersion.
- The delay differences may be caused by the dispersive properties of the waveguide material i.e. material dispersion and guidance effects within the fiber structure called waveguide dispersion.

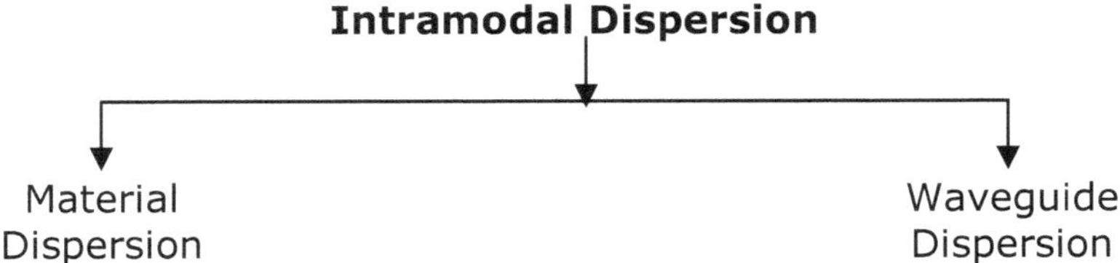

2.11.2 Material Dispersion

- Pulse broadening due to material dispersion results from the different group velocities of the various spectral components launched into the fiber from the optical source.
- It occurs when the phase velocity of a plane wave propagating in the dielectric medium varies non-linearly with wavelength and a material is said to exhibit material dispersion when the second differential of the refractive index with respect to wavelength is not zero.

- The pulse spread due to material dispersion may be obtained by considering the group delay τ_g in the optical fiber and is reciprocal of the group velocity v_g.

- Hence group delay.

  where,

$$\tau_g = \frac{d\beta}{d\omega} = \frac{1}{C}\left(n_1 - \lambda\frac{dn_1}{d\lambda}\right) \qquad \ldots (2.8)$$

$$n_1 \rightarrow \text{ Refractive index of the core material}$$

- Pulse delay τ_m due to material dispersion in a fiber of length is L is,

$$\tau_m = \frac{L}{C}\left(n_1 - \lambda\frac{dn_1}{d\lambda}\right) \qquad \ldots (2.9)$$

- Fig. 2.22 shows the variation of the material dispersion parameter M with wavelength for pure silica.

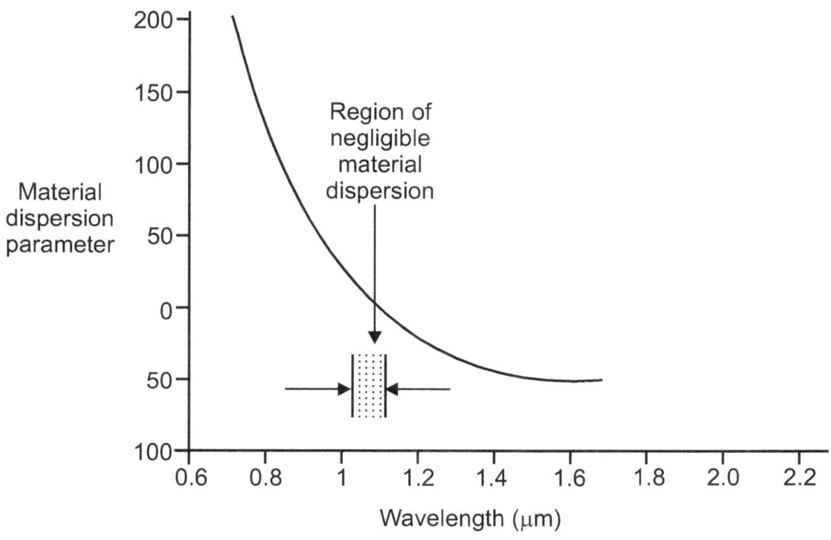

Fig. 2.22: Material Dispersion Parameter for Silica

- It is observed that the material dispersion tends to zero in the longer wavelength region around 1.3 μm for pure silica.

- This provides an additional incentive for operation at longer wavelengths where the material dispersion may be minimized.

- Also, the use of an injection LASER with a narrow spectral width rather than LED as optical source leads to a substantial reduction in pulse broadening due to material dispersion, even in the shorter wavelength region.

2.11.3 Waveguide Dispersion

- The waveguide dispersion of fiber also creates intramodal dispersion.

- This results from the variation in group velocity with wavelength for a particular mode.

2.11.4 Intermodal Dispersion   (W-14)

- **Pulse broadening due to intermodal dispersion** (sometimes referred as simply modal or mode dispersion) **results from the propagation delay differences between modes within a multimode fiber.**

- As the different modes which constitute a pulse in a multimode fiber travels along the channel at different group velocities, the pulse width at the output is dependent on the transmission times of the slowest and fastest modes.

- This dispersion mechanism creates the fundamental differences in the overall dispersion for three types of fiber as shown in Fig. 2.23.

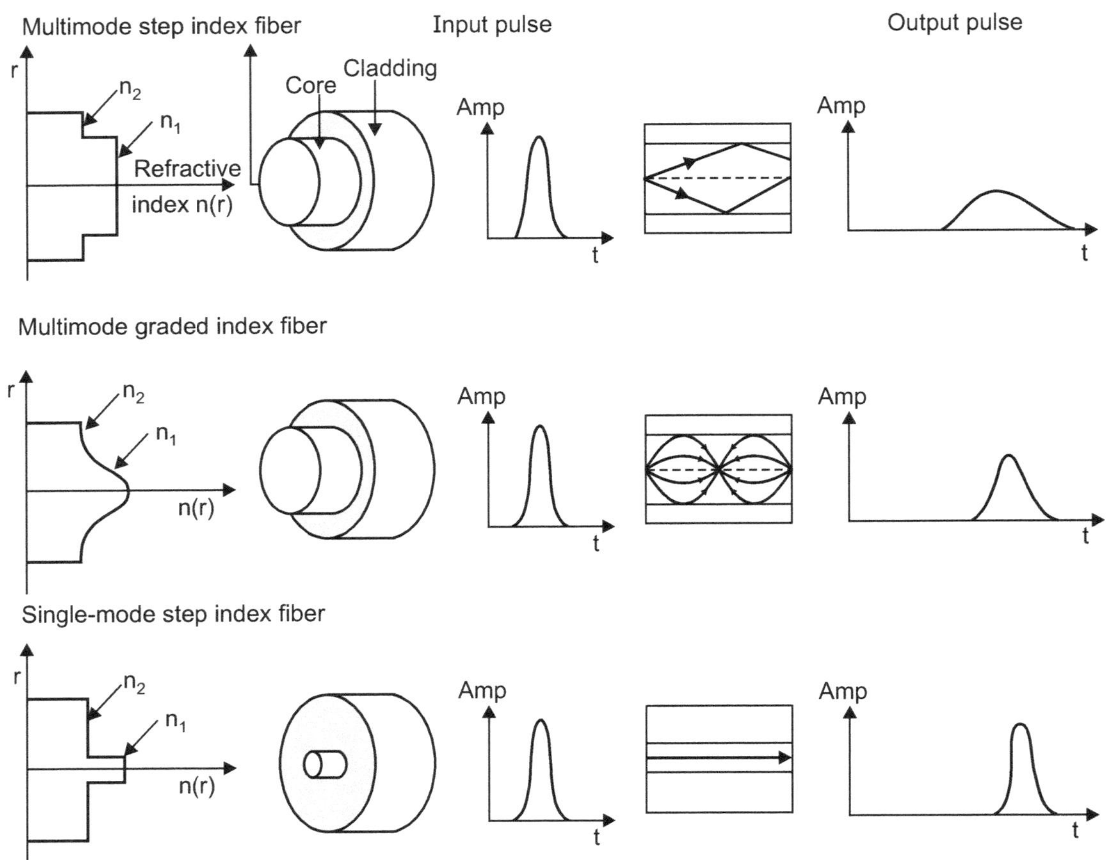

Fig. 2.23: Schematic Diagram Showing a Multimode Step Index Fiber, Multimode Graded Index Fiber and Single-Mode Step Index Fiber and Illustrating the Pulse Broadening due to Intermodel Dispersion in each Fiber Type

- Thus, multimode step index fibers exhibit a large amount of intermodal dispersion which gives the greatest pulse broadening.
- However, intermodal dispersion in multimode fibers may be reduced by adoption of an optimum refractive index profile.
- Hence, the overall pulse broadening in multimode graded index fibers is far less than that of multimode step index fibers. (Typically by a factor of 100).
- **Thus, graded index fibers used with a multimode source give a tremendous bandwidth advantage or multimode step index fibers.**

2.12 RADIATION LOSS

- **Radiation losses are caused mainly by small bends and links in the fiber.**
- Essentially, there are two types of bends: microbends and constant-radius bends.
- Microbending occurs as a result of differences in the thermal conduction rates between the core and the cladding material.
- A microbend is a miniature bend or geometric imperfection along the axis of the fiber and represents a discontinuity in the fiber where Rayleigh scattering can occur.
- Microbending losses generally contribute less than 20% of total attenuation in a fiber.
- Constant radius bends can be caused by excessive pressure and tension and generally occur when fibers are bent during handling or installation.

2.13 COUPLING LOSS

- **Coupling losses are caused by imperfect physical connections.**
- In fiber cables, coupling losses can occur at any of the following three types of optical junctions: Light source-to-fiber connections, fiber-to-fiber connections and fiber to photo detector connections.

- Junction losses are most often caused by one of the following alignment problems.
    1. Lateral misalignment
    2. Gap misalignment
    3. Angular misalignment and
    4. Imperfect surface finishes

2.13.1 Lateral Misalignment (Displacement)

- Fig. 2.24 (a) shows the lateral or axial displacement between two pieces of the adjoining fiber cables.

- The amount of loss can be from a couple tenths of a decibel to several decibels.

- This loss is generally negligible if the fiber axes are aligned to within 5% of the smaller fiber's diameter.

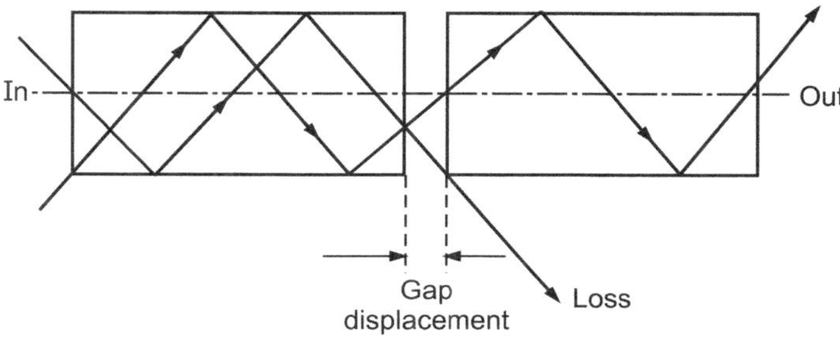

Fig. 2.24 (a): Lateral Misalignment

2.13.2 Gap Misalignment (Displacement)

- Gap misalignment or displacement is shown in Fig. 2.24 (b).

- When splices are made in optical fibers, the fibers should actually touch.

- The farther apart the fibers, the greater the loss of light.

- If two fibers are joined with a connector, the ends should not touch because the two ends rubbing against each other in the connector could cause damage to either or both fibers.

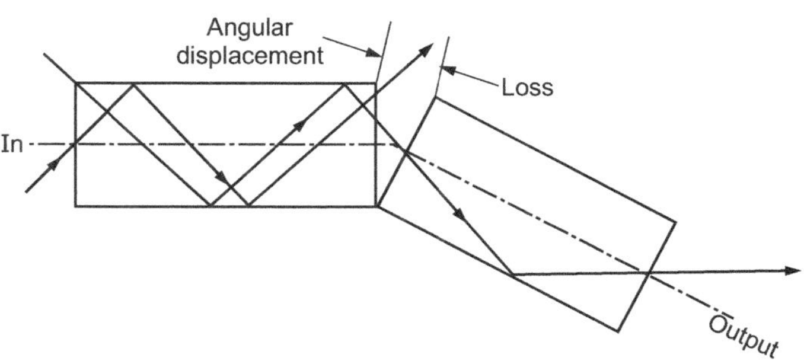

Fig. 2.24 (b): Gap Displacement

2.13.3 Angular Misalignment (Displacement)

- Angular displacement is shown in Fig. 2.24 (c).

- If the angular displacement is less than 3° off from perpendicular, the losses will typically be less than 0.5 dB.

Fig. 2.24 (c): Angular Displacement

2.13.4 Imperfect Surface Finish

- Imperfect surface finish is shown in Fig. 2.24 (d).

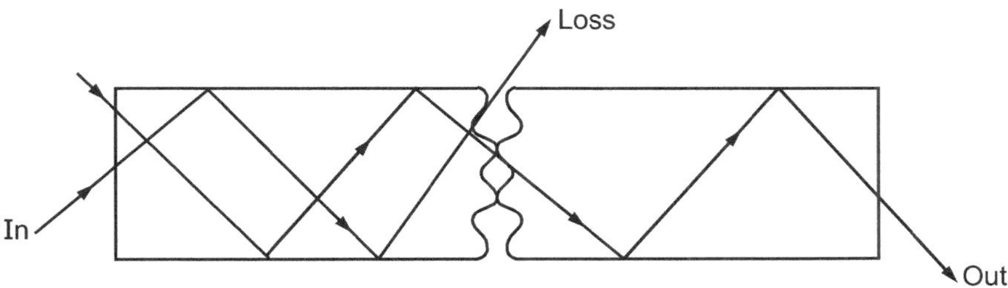

Fig. 2.24 (d): Imperfect Surface Finish

- The ends of the two adjoining fibers should be highly polished and fit together squarely.

- If the fiber ends are less than 3° off from perpendicular, the losses will typically be less than 0.5 dB.

2.14 OPTICAL TIME DOMAIN REFLECTOMETRY (OTDR)   (S-15)

QUESTION

1.   Draw and explain the block diagram of OTDR.   (8M)

- A measurement technique which is more sophisticated and which finds wide applications in both the laboratory and the field is the use of Optical Time Domain Reflectometry (OTDR). This technique is often called the backscatter measurement method.

- It provides measurement of the attenuation on an optical link down its entire length giving its information on the length dependence of the link loss.

- When the attenuation on the link varies with length, the averaged loss information is inadequate, OTDR also allows splice and connector losses to be evaluated as well as the rotation of any faults on link.

- **Block Diagram of OTDR:**

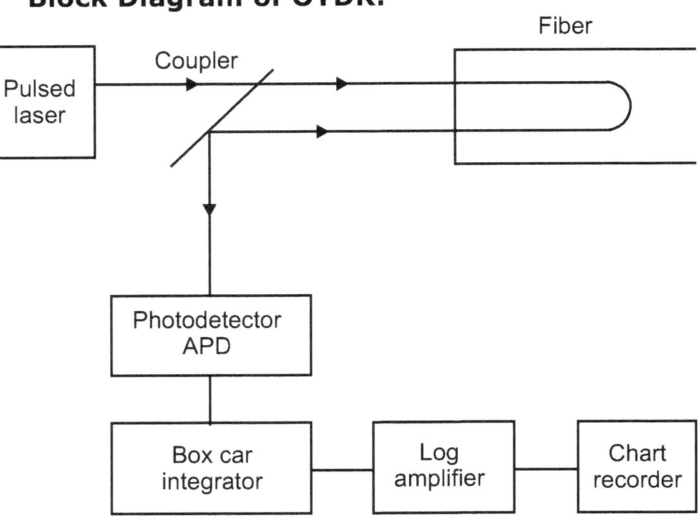

Fig. 2.25: Optical Time Domain Reflectometry (OTDR) or Backscatter Measurement Method

- It relies upon the measurement and analysis of the fraction of light which is reflected back within the fiber's numerical aperature due to Rayleigh scattering.

Working:

- A light pulse is launched into the fiber in the forward direction from an **injection laser** using either a **direction coupler** or a system external lenses with a beam splitter.

- The backscatter light is detected using an **avalanche photodiode** receiver which drives an **integrator** in order to improve the received signal to noise ratio. It also gives an arithmetic average over a number of measurements taken at one point within the fiber.

- The signal from the **integrator** is fed through a **logarithmic amplifier** and averaged measurements for successive points within the fiber are plotted on a chart recorder.

- This plot provides location dependent attenuation values which give an overall picture of the optical loss down the link.

- A possible backscatter plot shown in Fig. 2.26.

Output Description:

- Plot in Fig. 2.26 shows that the initial pulse caused by reflection and backscatter from the input coupler followed by a long tail caused by the distributed Rayleigh scattering from the input pulse as it travels down the link.

- Also shown in plot is a pulse corresponding to the discrete reflection from a fiber joint, as well as discontinuity due to excessive loss at a fiber imperfection or fault.

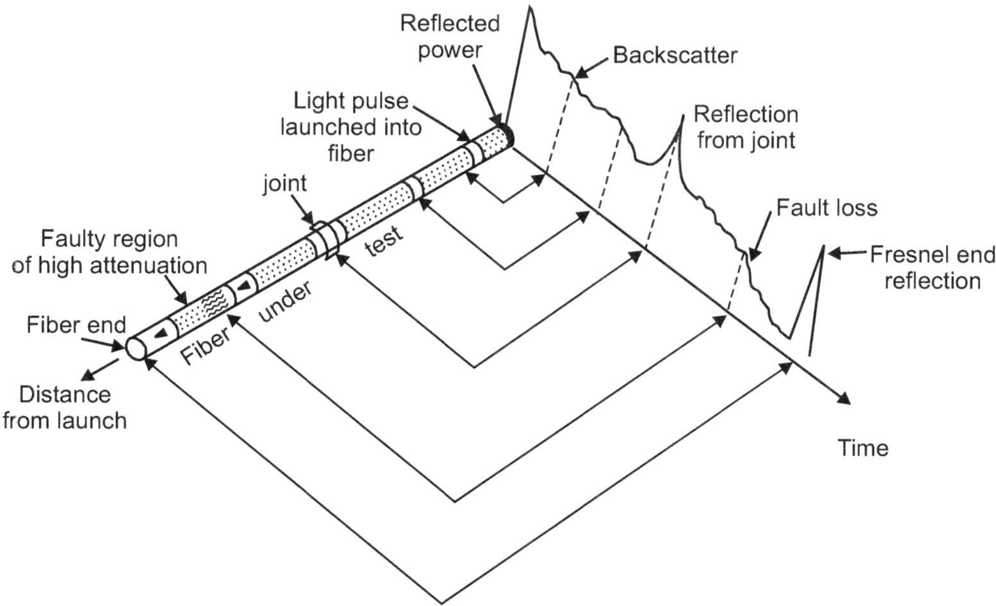

Fig. 2.26: Possible Backscatter Plot from a Fiber under Test

- The end of the fiber link is indicated by a pulse corresponding to the Fresnel reflection incurred at the output end face of the fiber.

- Such a plot yields the attenuation per unit length for the fiber by simply computing the slope of the curve over the required length.

- Also the location and insertion losses of joints and faults can be obtained from the power drop at their respective positions on link.

- Finally, the overall link length can be determined from the time difference between reflections from the fiber input and output end faces.

Important Points

- The predominant losses in optical fiber cable are:
  1. Adsorption loss
  2. Scattering loss
  3. Dispersion loss
  4. Radiation loss
  5. Coupling loss

- Attenuation means the loss of light energy as the light pulse travels from one end of cable to the other.

- Attenuation is directly proportional to the length of the cable.

- Material absorption is a loss mechanism related to the material composition and the fabrication process for the fiber, which results in dissipation of some of the transmitted optical power as heat in the waveguide.

- The absorption of the light may be:

  (i) Intrinsic: Caused by the interaction with one or more of the components of glass.

  (ii) Extrinsic: Caused by impurities within the glass.

- A pure silicate glass has small intrinsic absorption due to its basic material structure in the near infra-red region.

- Extrinsic absorption result from the presence of impurities.

- Scattering loss is one type of attenuation.

- Scattering is the loss of optical energy due to imperfections in the fiber and from the basic structure of the fiber.

- **Scattering Losses**

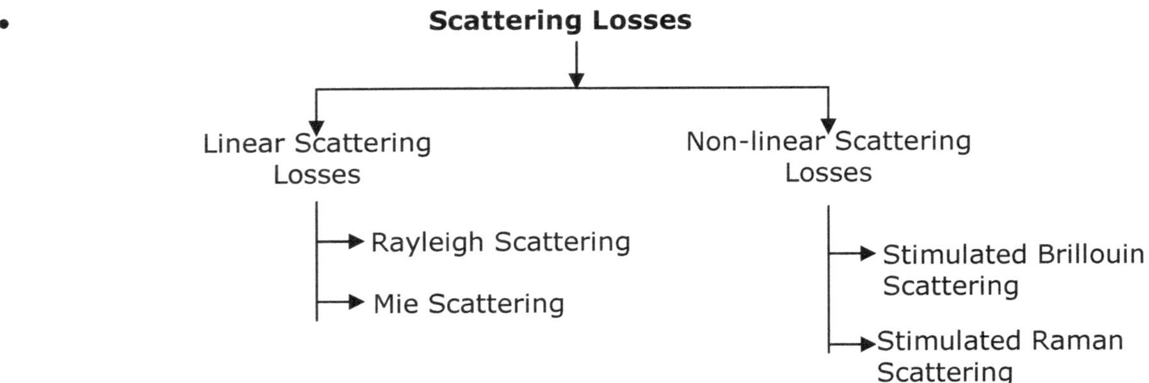

- Linear scattering mechanisms cause the transfer of some or all of the optical power contained within one propagating mode to be transferred linearly into different mode.
- Linear scattering categorized in two types:

  (i) Rayleigh scattering.   (ii) Mie scattering.

- Rayleigh scattering is the dominant intrinsic loss mechanism in the low absorption window between the ultraviolet and infrared absorption tails.
- The subsequent scattering due to the density fluctuations, which is almost in all directions, produces an attenuation proportional to $\frac{1}{\lambda^4}$ following the Rayleigh scattering formula.
- The scattering created by such inhomogeneities is mainly in forward direction, is called Mie scattering.
- Optical wavelength do not always behave as completely linear channels whose increase in output optical power is directly proportional to the input optical power.
- There are two important types of non-linear scattering within optical fibers are:

  (i) Stimulated Brillouin and   (ii) Raman scattering

- Stimulated Brillouin Scattering (SBS) regarded as the modulation of light through thermal molecular vibrations within the fiber.
- **Stimulated Raman Scattering (SRS)** is similar to SBS except that a high frequency optical photon rather than an acoustic photon is generated in the scattering process.
- Dispersion of light is the distortion of light pulse and it is broadening of light pulse and merging of pulse into another pulse.
- Digital modulation used for optical fiber, then dispersion mechanisms within the fiber cause broadening of the transmitted light pulses as they travel along the channel.
- **Dispersion**

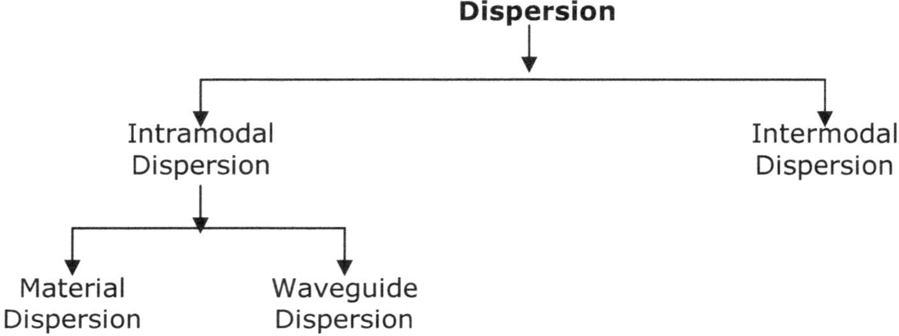

- Intramodal or chromatic dispersion may occur in all types of optical fiber and results from the finite spectral linewidth of the optical source.

- 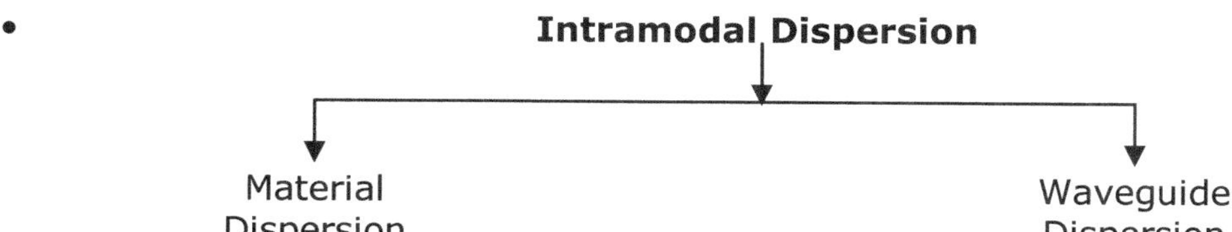

- Pulse broadening due to material dispersion results from the different group velocities of the various spectral components launched into the fiber from the optical source.

- The waveguide dispersion of fiber also creates intramodal dispersion.

- This results from the variation in group velocity with wavelength for a particular mode.

- **Pulse broadening due to intermodal dispersion** (sometimes referred as simply modal or mode dispersion) **results from the propagation delay differences between modes within a multimode fiber.**

- Radiation losses are caused mainly by small bends and links in the fiber.

- Coupling losses are caused by imperfect physical connections.

- Junction losses are most often caused by one of the following alignment problems.

  1. Lateral misalignment
  2. Gap misalignment
  3. Angular misalignment and
  4. Imperfect surface finishes

- Splicing is the technique to join the two fiber cables at the bend.

- **A permanent joint formed between two individual optical fibers in the field or factory is known as a fiber splice.**

- Splices are divided into two categories depending upon the splicing technique utilized.

  (i) Fusion splicing or Welding

  (ii) Mechanical Splicing.

- The fusion splicing of single fibers, two prepared fiber ends heated to their fusion point with the application of sufficient axial pressure between the two optical fibers.

- A simple method utilizes a V-groove into which the two prepared fiber ends are pressed.

- Fiber connectors are special mechanical assemblies that allow fiber optic cable to be connected to one another.

- Fiber optic connectors are the optical equivalent of electrical plugs and sockets.

- A measurement technique which is more sophisticated and which finds wide applications in both the laboratory and the field is the use of Optical Time Domain Reflectometry (OTDR). This technique is often called the backscatter measurement method.

Practice Questions

1. Describe linear scattering losses in optical fibers.

   (i) Rayleigh scattering

   (ii) Mie scattering

2. Compare Stimulated Brillouin and Stimulated Raman Scattering in optical fibers and indicate the way in which they may be avoided in optical fiber communications.

3. Explain the reasons for pulse broadening due to material dispersion in optical fiber

4. Describe intrinsic absorption in fiber optic.

5. Describe extrinsic absorption in fiber optic.

6. List scattering losses and explain linear scattering loss.

7. State and explain Rayleigh scattering.

8. Explain Stimulated Brillouin Scattering and Stimulated Raman scattering.

9. Draw block diagram of OTDR and explain.

10. Describe link and power budget.

11. Explain attenuation and absorption loss.

12. Explain intramodal and intermodal dispersion.

13. Define reflection and refraction.

14. Explain Snell's law and critical angle.

15. Explain coupling loss.

16. Give description of elastic tube splice with neat diagram.

Optical Network

Syllabus

3.1   Optical Network Components Use and Features : Amplifiers, Splitter, Optical Switches.

3.2   WDM : Basic Concept, Features

3.3   SONET / SDH : Architecture and Hierarchy

3.4   Ethernet Standards of Optical Network Features : IEEE 802.3j, 802.3y, 802.3z

About this Chapter

At the end of this chapter students will be able to:

- Describe working principle of the optical network components.
- Explain the concept of WDM.
- Explain the architecture of SONET/SDH.
- Describe the given type of Ethernet standard.

3.1 INTRODUCTION

- To amplify an optical signal with a conventional repeater, one performs photon to electron conversion, electrical amplification, retiming, pulse shaping and then electron to photon conversion.

- Optical amplifier contains three types as SOAs, DFAs and EDFAs.

- Optical network components also uses splitters which are the passive optical device that enables a light signal on an optical fiber to be distributed among two or more fibers.

- Optical switches are of two types :

  (i) Circuit switch   (ii) Packet switch

- Optical Ethernet is designed to send standard - well known 10/100 - Mb/s, 1 Gb/s and 10 Gb/s Ethernet frames directly over optical fibers.

- Optical Ethernet offers reduced network complexity.

3.2 OPTICAL NETWORK ELEMENTS

- A set of optical network elements connected by optical fiber links, able to provide functionality of transport, multiplexing, switching, management, supervision and survivability of optical channels carrying client signals.

- Optical network components are as follows :

3.2.1 Amplifiers (Optical Amplifiers)

- As the name suggests, optical amplifiers operate solely in the optical domain with no interconversion of photons to electrons.

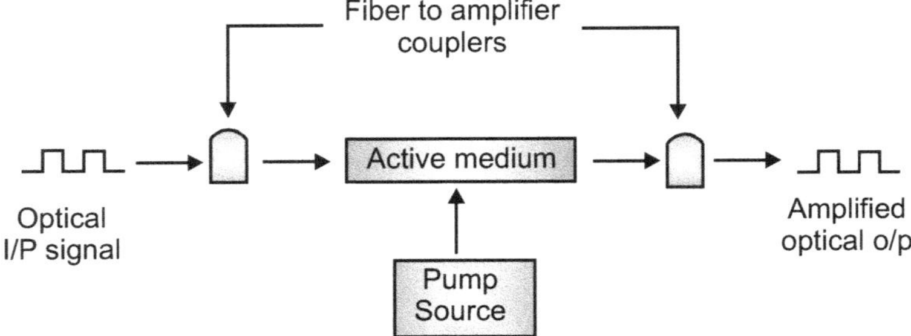

Fig. 3.1: Operation of Optical Amplifiers

Working:

- The operation of optical amplifier is as shown in Fig. 3.1. Here, the absorption of energy in device is supplied from an external source called **pump**.
- The pump is used to supply energy to electrons in active medium, that raises them to higher energy levels which produces a population inversion.
- Optical input signal (incoming signal photon) will trigger these excited electrons to drop to lower levels through a stimulated-emission process.
- The three main optical amplifier types can be classified as semiconductor optical amplifier (SOA$_3$), **D**oped-**F**iber **A**mplifier (DFAs) and **R**aman **A**mplifier.
- **"An optical amplifier is a device which amplifies the optical signal directly without ever changing it to electricity. The light itself is amplified".**

3.2.1.1 Applications of Amplifier

1.

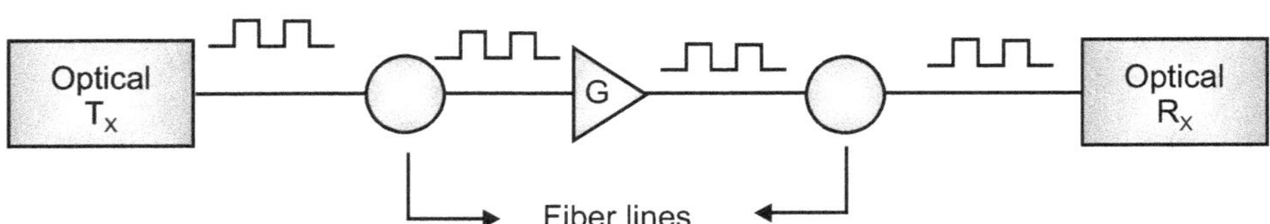

(a) Inline Amplifier to Increase Transmission Distance

2.

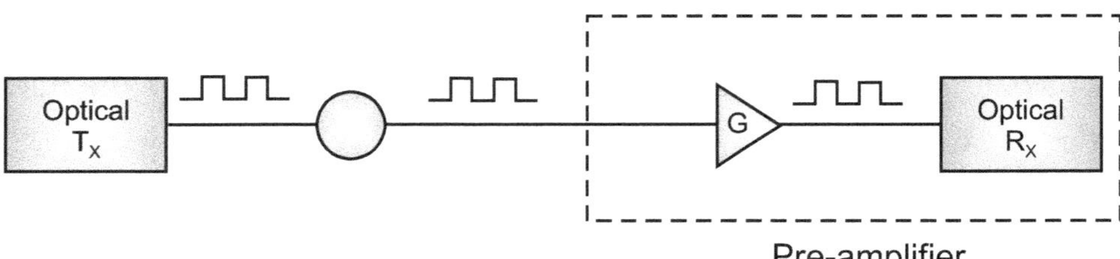

(b) Preamplifier to Improve Receiver Sensitivity

3.

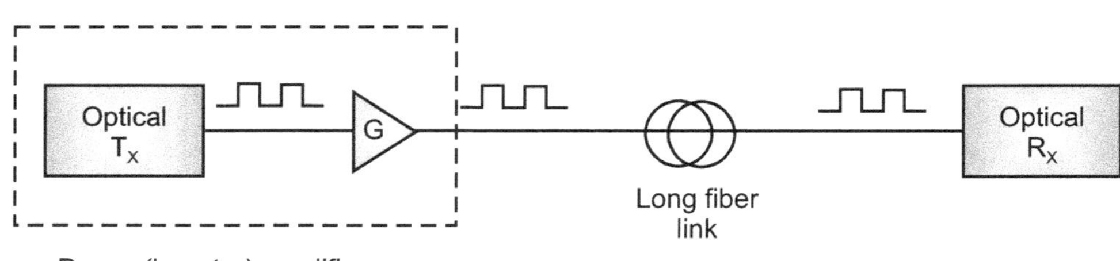

(c) Booster of Transmitted Power

4.

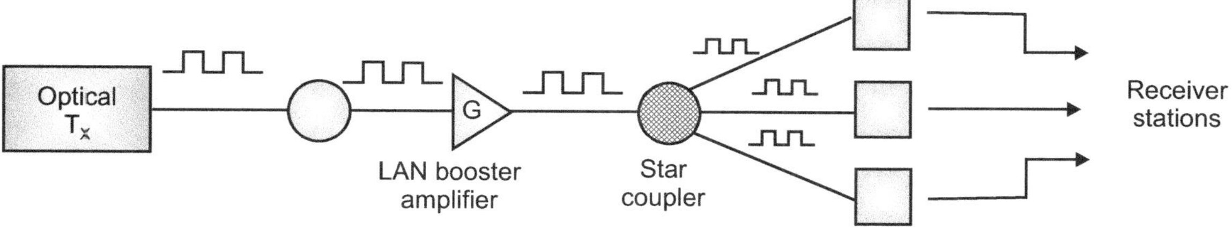

(d) Booster of Signal Level in LAN

Fig. 3.2

3.2.1.1.1 Advantages of SOA

1. Small size.
2. Bidirectional transmission.
3. Smaller output power then EDFA.
4. Less expensive than EDFA.

3.2.1.1.2 Disadvantages of SOA

1. Lower gain then EDFA.
2. High non-linearity.
3. Polarization dependence.
4. Higher noise then EDFA.

3.2.1.1.3 Advantages of Raman Amplifier

1. Can be used to "extend" EDFAs.
2. Compatible with installed SM fiber.
3. Variable wavelength amplification is possible.

3.2.1.1.4 Disadvantages of Raman Amplifier

1. High pump power requirements.
2. Needs sophisticated gain control.
3. Noise.

3.2.1.1.5 Advantages of EDFA

1. Low polarization sensitivity.
2. Low noise
3. High gain.

3.2.1.1.6 Disadvantages of EDFA

1. Large size.
2. High pump power consumption.

3.2.2 Optical Splitters

Definition:

- **"A fiber optic splitter is a passive optical device that enables a light signal on an optical fiber to be distributed among two or more fibers.**

Working:

- When the light beam transmitted in a network needs to be divided into two or more light beams, fiber optic splitters are used.

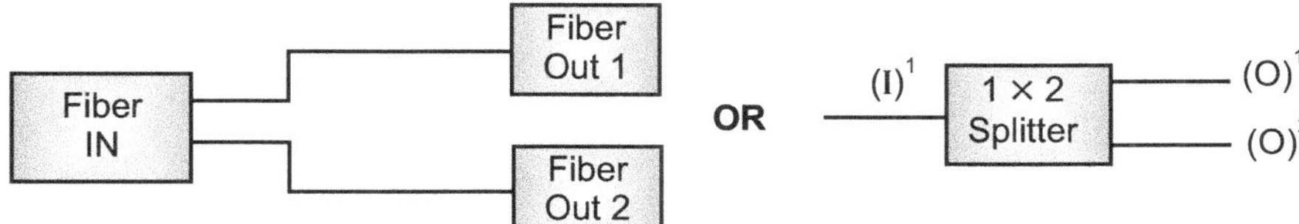

Fig. 3.3 : 1 × 2 Optical Splitter

- When the light signal is transmitted in a single mode fiber, the light energy cannot entirely concentrate in the fiber core.

- If two fibers are close enough to each other, the transmitting light in an optical fiber can enter into another optical fiber.

- Splitter does not generate power nor require power. Hence, it is a passive device. Also, splitter does not contain any electronic components.

- It is a simple device, also known as **beam splitter**.
- As shown in Fig. 3.3, splitters has many input and output terminals, especially applicable to passive optical networks (GPON, BPON, FTTX, EPON etc.).
- The fiber optic splitters can be divided into two types: **Fused Biconical Taper (FBT)** and **Planar Lightwave Circuit (PLC) Splitter**.
- The **FBT** splitters are the most commonly used splitters. FBT splitters are widely accepted and used in passive networks, especially for instances where the split configuration is smaller (1×2, 1×4, 2×2 etc.).

3.2.2.1 Advantages of FBT Splitter

1. Can make unequal splitting ratio.
2. Low cost.
3. Easy availability of raw materials.

3.2.2.2 Disadvantages of FBT Splitter

1. Poor uniformity.
2. Insertion loss changes according to the change in temperature variation.
3. Loss sensitive wavelength.
4. Cannot ensure uniform spectroscopic.

3.2.2.3 Advantages of PLC Splitter

1. Small volume.
2. Low cost.
3. Compact structure.
4. Spectral uniformity.

3.2.2.4 Disadvantages of PLC Splitter

1. High technical threshold.
2. Device fabrication process complexity.

3.2.3 Optical Switches

- The function of an optical switch is to switch data between different parts of network.
- An optical switch performs various digital logic operations which allows signals to switch from one state to another. Therefore, larger arrays of optical switches can switch signals from one port to another.
- As with electronic switching, optical switching can be classified into two types:

  (i) Circuit switching   **(ii) Packet switching**
- Both circuit and packet switching technique are used in high capacity networks.
- In circuit-switched networking, the connection must be set-up between the transmitter and receiver, before initiating the transfer of data.

3.2.3.1 Optical Circuit-Switched Networks

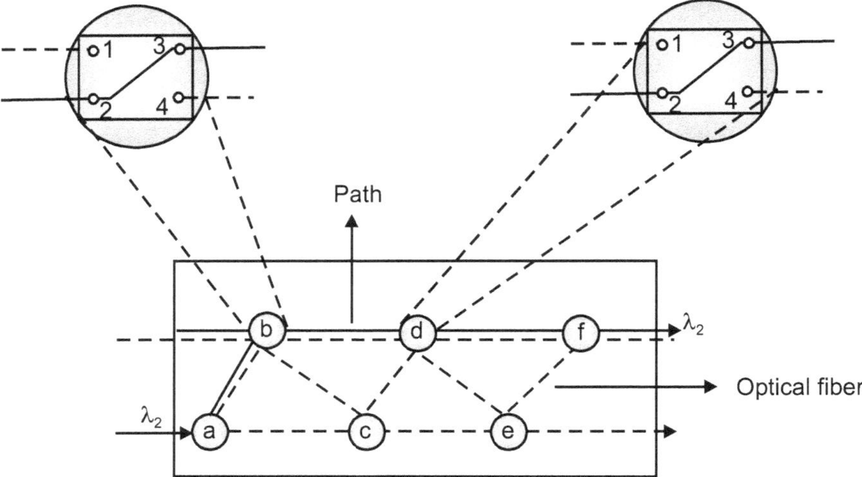

Fig. 3.4: Optical Circuit Switch Network

Working:

- In circuit-switched networking, a connection is made using available network resources for the full deviation of transmission of message.
- Once the complete message, is successfully transmitted then the connection is removed.
- A circuit-switched environment requires, that on end-to-end circuit be made before the actual transmission can take place.
- A request signal must, however, travel from the source to the destination and it should also be acknowledged before the transmission begins.
- As shown in Fig. 3.4, six optical nodes (a to f) are interconnected and a requested logical connection or path for optical signal wavelength, λ_1, is established producing a circuit path through network nodes a, b, d, and f.
- Optical nodes of an OCS (Optical Switched Network), network contain optical switches where large multiport optical switches are used to establish connections between the desired input and output ports.
- Fig. 3.4 illustrates this functionality at optical nodes b and d. For example, 2×2 optical switch at node b enables cross-connection using ports 2 and 3, whereas node d employs a direct connection for ports 1 and 3.
- Hence, a logical path or an optical circuit from node a to f is created for the light path using signal wavelength λ_1.
- A disadvantage of OCS is that it cannot efficiently handle bursty traffic. In such traffic conditions the data is sent in optical bursts of different lengths and therefore the resources cannot be readily assigned.

3.2.3.2 Optical Packet-Switched Networks (OPS)

- In OPS, data is transported entirely in the optical domain without intermediate optoelectrical conversion.
- OPS, performs the four basic functions such as **routing**, **forwarding**, **switching** and **buffering**.
- **Routing:** It provides network connectivity information often through preallocated routing tables.
- **Forwarding:** It defines the output for each incoming packet based on routing table.
- **Switching:** The switch directs each packet to the correct output defined by the forwarding process.
- **Buffering:** It provides, data storage for packets to resolve any contention problems which may occur during packet transmission.

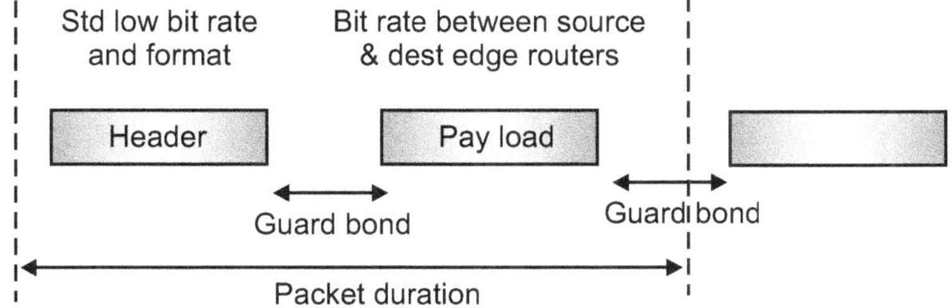

Fig. 3.5: Optical Packet-Switched Network Packet Format

- As shown in Fig. 3,5, header and payload requires a guard band to ensure the data is not overwritten.
- The header points to an entry in a lookup table that specifies to where the packet should be forwarded.

3.2.3.2.1 Advantages of Optical Switching

1. Not affected due to interference as found in EM wave propagation.
2. Provides better security.
3. More efficient and more faster.
4. Reduced size of node.
5. Very much flexible utilization of network.

3.2.3.2.2 Disadvantages of Optical Switching

1. Higher control complexity.
2. Requires more effort for packet recording.
3. High wavelength consumption.
4. Large node size.

3.3 WAVELENGTH DIVISION MULTIPLEXING (WDM)

- WDM involves the transmission of a number of different peak wavelength optical signals in parallel on a single optical fiber.
- The technology is illustrated in Fig. 3.6, where a conventional i.e. single nominal wavelength optical fiber communication system is shown together with a duplex and also a multiplex fiber communication system.
- It is the latter WDM operation which has generated particular interest within telecommunication.

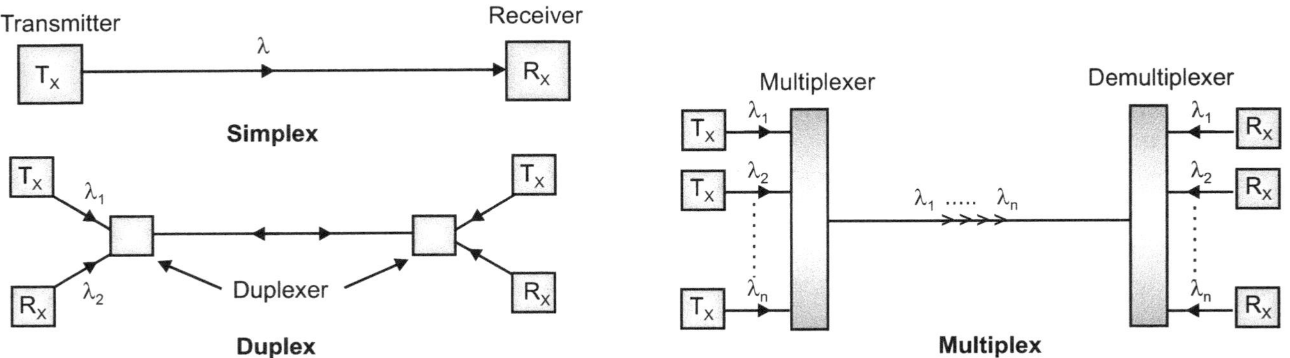

Fig. 3.6: OFC Operating Modes Showing WDM

Operational Principle of WDM:

- Fig. 3.6 shows the implementation of passive and active components in a typical WDM link containing various types of optical amplifiers.
- At the transmitting end there are several independently modulated light sources, each emitting signals at a unique wavelength. Here, a multiplexer is needed to combine these optical outputs into a continuous spectrum of signals and couple them onto a single fiber.
- At the receiving end, a demultiplexer is used to separate the optical signals into appropriate detection channels for signal processing.

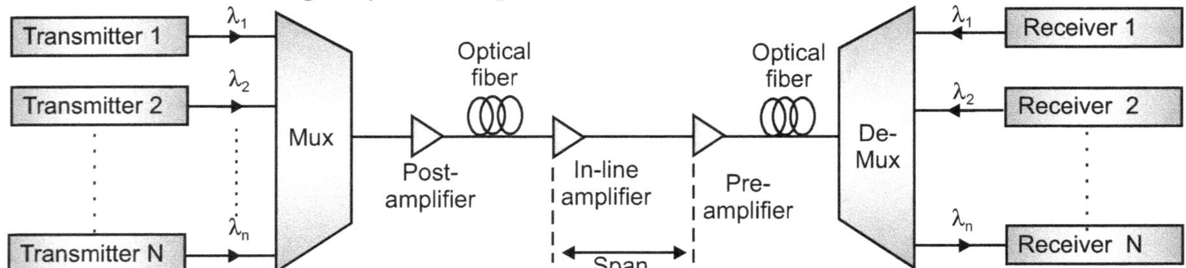

Fig. 3.7: Block Schematic of WDM

3.3.1 Features of WDM

1. **Capacity Upgrade:** WDM system can increase the capacity of a fiber network.
2. **Transparency:** In WDM, each optical channel can carry any transmission format with WDM, any type of information, analog or digital can be sent simultaneously over the same fiber.
3. **Wavelength Routing:** In addition to using multiple wavelengths to increase link capacity, the use of wavelengths sensitive optical routing devices makes it possible to use wavelength as another dimension in designing communication networks and switches.

3.3.2 Advantages of WDM

1. Provides higher BW.
2. High security.
3. Easier to reconfigure.
4. Full duplex transmission is possible.
5. Optical components are more realiable.

3.3.3 Disadvantages of WDM

1. Cost of system increases with addition of optical components.
2. Signals cannot be very close.
3. Light wave carrying WDM are limited to 2-point circuit.
4. Inefficiency in bandwidth utilization.
5. Difficulty in wavelength tuning.

3.4 SONET / SDH

- The **Synchronous Optical Network** (SONET) is a standard for optical telecommunications transport established by **ANSI** in the **1980's**.
- The **Synchronous Digital Hierarchy** (SDH), was formulated by the **ITU** standards body.
- Three basic devices used in SONET system are as follows:

(i) Synchronous Transport Signal (STS) MUX/Demux:

- It either multiplexes signal from multiple sources into a STS signal or demultiplexes an STS signal into different destination signals.

(ii) Regenerator:

- It is a repeater that takes a received optical signal and regenerates it. It functions in data link layer.

(iii) Add/Drop MUX:

- It adds signals coming from different sources into a given path or remove a desired signal from a path and redirects it without demultiplexing the entire signal.

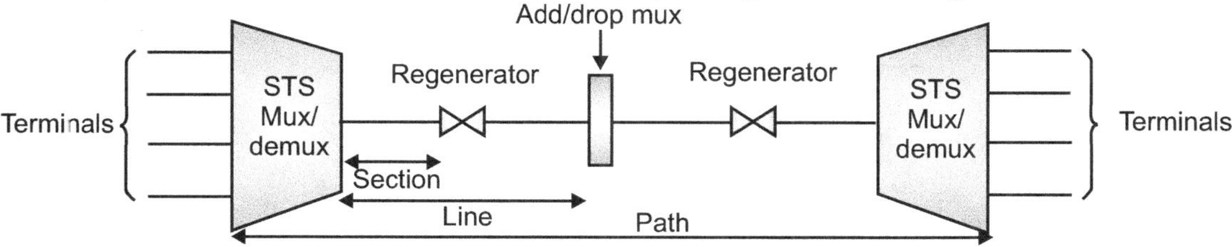

Fig. 3.8: Elements of SONET

Functional Layers:

- The SONET standard includes four functional layers as: Photonic layer, the section layer, the line and path layer. They correspond to both physical and data link layer.

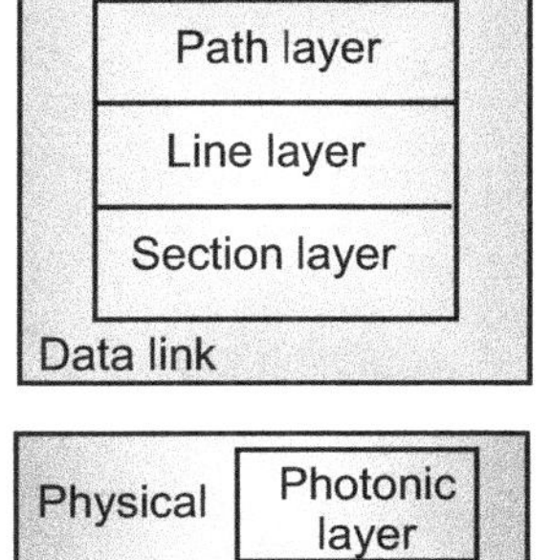

Fig. 3.9: Functional Layers of SONET

Description of Layers:

1. Path Layer:

- It is responsible for the movement of signal from its output source to its optical destination.
- STS MUX/demux provides path layer function.

2. Line Layer:

- It is responsible for the movement of signal across the physical line.
- STS MUX/demux and add/drop MUX provides line layer functions.

3. Section Layer:

- It is responsible for the movement of signal across a physical section.
- Each device of network provides section layer functions.

4. Photonic Layer:

- It corresponds to the physical layer of the OSI model.
- It includes physical specifications for the optical fiber channel. (Presence of light = 14 absence of light = θ).

3.4.1 Advantages of SONET

1. Transits data to large distance.
2. Low electromagnetic interference.
3. High data rates.
4. Large bandwidth.
5. Increased reliability and restoration over electrical systems.
6. Capability to build optical interconnects between carriers.
7. Reduced network complexity and cost.
8. Very high efficiency.
9. Standard optical interference.
10. De-multiplexing is easy.
11. Remote operation capabilities.
12. Out of band management system.

3.4.2 Disadvantages of SONET

1. No interoperable standard.
2. Bandwidth efficiency is a problem at higher capacity.
3. More overhead is required.
4. Tributary services require SONET MUX services.

3.5 ETHERNET STANDARD

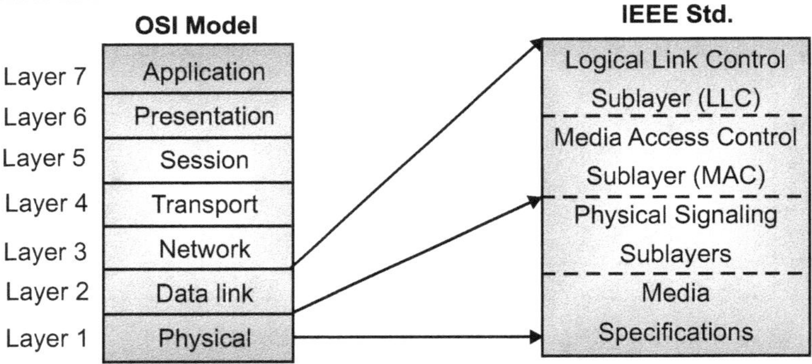

Fig. 3.10: Ethernet Standards

- Ethernet is a communication standard that was developed in the early 80's to network computers and other devices in a local environment such as home or building.
- This local environment is defined as a **LAN** (Local Area Network) and it connects multiple devices so that they can create, store and share information with others in the location.
- Ethernet is a wired system that started with using coaxial cable and has successfully progressed to now using twisted pair copper wiring and fiber optic wiring.
- In 1983, Ethernet was standardized into the standard IEEE 802.3 by the **Institute of Electrical and Electronic Engineers** (IEEE).
- The standard defined the physical layer and **MAC** (media access control) portion of the "data link" layer of wired Ethernet.
- These two layers are defined as the first two layers in the **OSI** (Open Systems Interconnection) model.

802.11	Concept
802.11 a	• Specifies a PHY that operates in the **5 GHz** band in the US-initially 5.15-5.35 AND 5.725-5.85. Since expanded to additional frequencies. • Uses Orthogonal Frequency Division Multiplexing (OFMD). • Enhanced data speed to 54 Mbps.
802.11 b	• Enhancement to 802.11 that added higher data rate modes to the DSSS already defined in the original 802.11 standard. • Boosted data speed to 11 Mbps. • 22 MHz BW yields 3 non-overlaping channels in the frequency range of 2.400 GHz to 2.4835 GHz.
802.11 d	• Enhancement to 802.11 a and 802.11 b that allows for **global roaming**. • Particulars can be set at MAC layer.
802.11e	• Enhancement to 802.11 includes quality of service (QoS) features. • Facilitates prioritization of data, voice and video transmissions.
802.11 g	• Extends the maximum data rate of WLAN devices that operates in the 2.4 GHz band in a fashion that permits interoperation with 802.11 b devices. • Uses OFDM modulation (orthogonal FDM). • Operates at upto 54 Mbps, with fall-back speed that includes the "b" speeds.
802.11 h	• Enhancement to 802.11a that resolves interference issues. • Dynamic Frequency Selection (DFS). • Transmit Power Control (TPC).
802.11 i	• Enhancement to 802.11 that offers additional security for **WLAN** applications. • Defines more robust encryption, authentication and key exchange.
802.11 j	• Japanese regulatory extension to 802.11 a specification. • Frequency range 4.9 GHz to 5.0 GHz.
802.11 k	• Radio resource measurements for network using 802.11 family specifications.
802.11 m	• Maintenance of 802.11 family specifications. • Corrections and amendments to existing documentation.
802.11 n	• Higher-speed standards. • Top speed claimed of 108,240 and 350 + MHz.
802.11 x	• Mis-used "generic" term for 802.11 family specifications.
802.11 z	• Ethernet standard that transmitted at 1 gigabit per second. It connects PCs and servers in local networks and is commonly employed along with a mix of 10/100 Mbps devices. • It transmits full duplex from point to point using Ethernet switches and half duplex in a shared Ethernet hub environment.

Important Points

- Optical network contains the component such as amplifier, splitter and switches.

- Amplifiers are having three types as:

  (i) Semiconductor Optical Amplifier (SOA)   (ii) Earth Doped Fiber Amplifier (EDFA)

  (iii) Raman Amplifier.

- When the light beam transmitted in a network, needs to be divided into two or more light beams fiber optic **splitters** are used.

- Fiber optic splitters are divided into two types :

  (i) Fused Biconical Taper (FBT) Splitter   (ii) Planar Lightwave Circuit (PLC) Splitter.

- Optical switch, switches the data between different parts of network.

- Optical switching is again classified into two groups :

  (i) Circuit switching   (ii) Packet switching.

- **WDM** is a technique in fiber optic transmission that enables the use of multiple light wavelengths to send data over the same medium.

- **SONET** is a standard for optical telecommunication transport established by ANSI in 1980's.

- **SONET** uses three basic devices as follows :

  (i) Synchronous Transport Signal (STS) MUX/demux

  (ii) Regenerator   (ii) Add/Drop MUX

- **Ethernet** is a communication standard that was developed in early 80's to network computers and other devices in a local environment such as home or building.

Practice Questions

1. Explain amplifiers used in optical networking.
2. Give idea about splitters in networking.
3. State optical switches used in optical networking.
4. Give basic concepts and features of WDM.
5. Give advantages and disadvantages of WDM.
6. Explain working principle of WDM.
7. Explain architecture of SONET/SDH.
8. Give Ethernet standard of optical networks.
9. Give advantages and disadvantages of SONET/SDH.
10. Give applications of amplifiers in optical network.

Chapter 4

Overview of Satellite Systems

Weightage of Marks = 12, Teaching Hours = 08

Syllabus

4.1 Working Principle, Concepts and Basic Components of Satellite System : Earth Segment, Space Segment, Active and Passive Satellite, Geostationary and Geosynchronous Satellites

4.2 Frequency Allocations for Satellite Services, Uplink and Downlink Frequency, Satellite Frequency Bands

4.3 Basic Terminologies used in Satellite Communication: Latitude, Longitude, Look Angle, Elevation Angle, Station Keeping, Propagation Delay Time Velocity, Look Angle and Footprint

4.4 Communication Satellite Orbits and its Types: LEO, MEO, Elliptical Orbit and GEO, Parameters and Characteristics of Various Orbits

4.5 Kepler's law, Apogee and Perigee Heights, Orbit Perturbations, Effects of a Non-spherical Earth, Atmospheric Drag, Effect of Eclipse on Satellite Motion

About this Chapter

At the end of this chapter students will be able to:
- Describe with sketches the working principles of the given type of satellite.
- Explain with sketches the given terms related to satellite and orbit.
- Explain the parameters with respect to the given type of satellite orbit.
- Explain Kepler's law of planetary motion with respect to the given criteria.

4.1 INTRODUCTION TO SATELLITE

- The reality of instant communication of video, audio and data signals to or from virtually any place on earth.
- Acting as space-based signal relay stations, satellites overcome the difficulties and limitations in broadcasting beyond visible horizon.
- A satellite system represents a careful and finely integration of electronics, mechanical structure, rocketry and antenna design supported by a system of earth-based ground stations, computers and radar for tracking position precisely.
- Different orbits provide a selection of coverage of the earths surface with orbital times ranging from 90 minutes to 24 hours.
- Using frequencies in the gigahertz range, satellites are effective but have very large path losses from earth to satellite and satellite to earth.
- A complete system requires high gain antennas, sensitive and low noise front end receiver stages and careful planning.
- A communication satellite is a **microwave repeater.**
- It receives energy beamed up at it by an earth station, amplifies it and return it back to the earth.
- The frequency difference between the uplink and downlink prevents the interference between the signals.

- The satellite is POSITIONED at about **36,000 kms above the earth,** where it experiences very less gravitational pull of the earth.

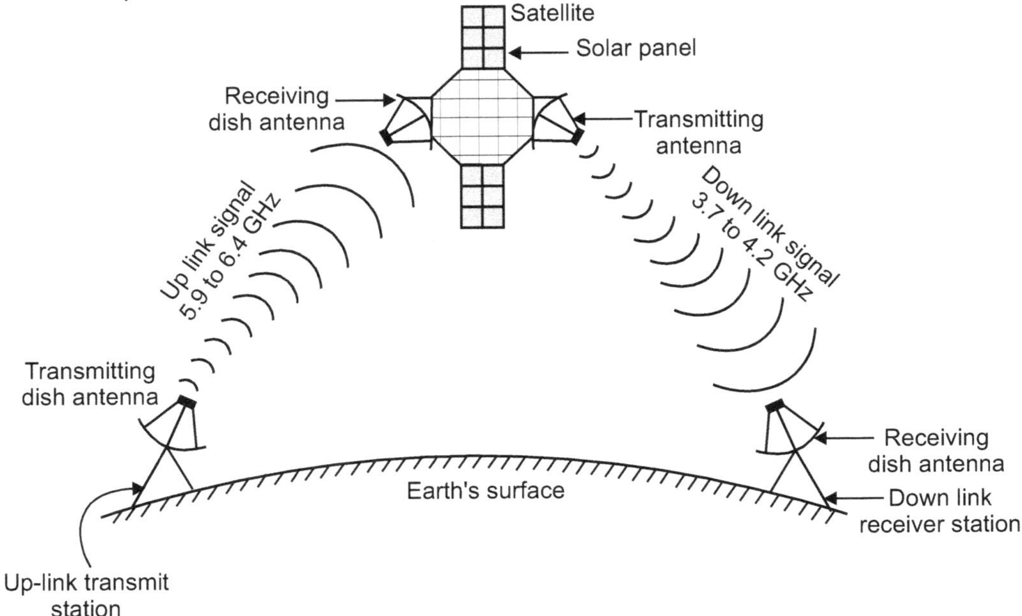

Fig. 4.1: Signal Paths in a Satellite Communication

- The information transmitted through earth station to the satellite.

- Thus, the satellite is large repeater in space that receives the modulated RF carrier in its uplink frequency spectrum.

- Received signal at satellite amplifies and retransmits them back to earth in down link frequency spectrum.

- The signal at receiving earth station is processed to get original information.

4.2 IMPORTANCE OF SATELLITE COMMUNICATION SYSTEM 

- The process of sending information around the world has been revolutionized by the advent of satellite.

- The feasibility of wideband communication over long distances has been demonstrated by the Telstar, Syncom and early Bird satellites.

- A good quality service can be continuously ensured by a global operational satellite system.

- Whereas communication using earth station transmitters covers very small area and to cover large area many transmitters has to be installed on earth surface.

- Satellites are installed in space near about 36,000 km from earth surface so that satellite covers very large area of earth surface.

- A satellite system should cater for carrying different types of signals which include telephony, data, telegraphy and facsimile. While television signals can be transmitted separately.

  For example,

  INTELSAT-I, INTELSAT-II, INTELSAT-III, INTELSAT-IV, INTELSAT-IVA.

- Also the telecommunication traffic is doubling every three years. Once in space, the capacity of satellite cannot be increased. At the same time, earth-station operators book the circuit only as traffic grows, as retal have to be paid for circuits on a permonth basis. INTELSAT has been able to meet this requirement.

- All earth stations have been able to obtain additional circuits immediately on placing an order for the same.

4.3 CONCEPT OF ORBIT AND ITS TYPES 

- An artificial satellite needs to have orbit, the path in which it travel around the earth.
- Most of satellites mentioned thus far are called orbital satellites, which are non-synchronous.
- Satellites which are not stationary with respect to earth are called non-synchronous satellites.
- A satellite remains in orbit because the centrifugal force caused by its rotation around the earth is counter-balanced by earth's gravitational pull.
- Non-synchronous satellites rotate around the earth in an elliptical or circular pattern, so there are two types of orbit.
  1. Circular orbit.
  2. Elliptical orbit
- Satellite orbits are shown in Fig. 4.2 (a) and 4.2 (b).

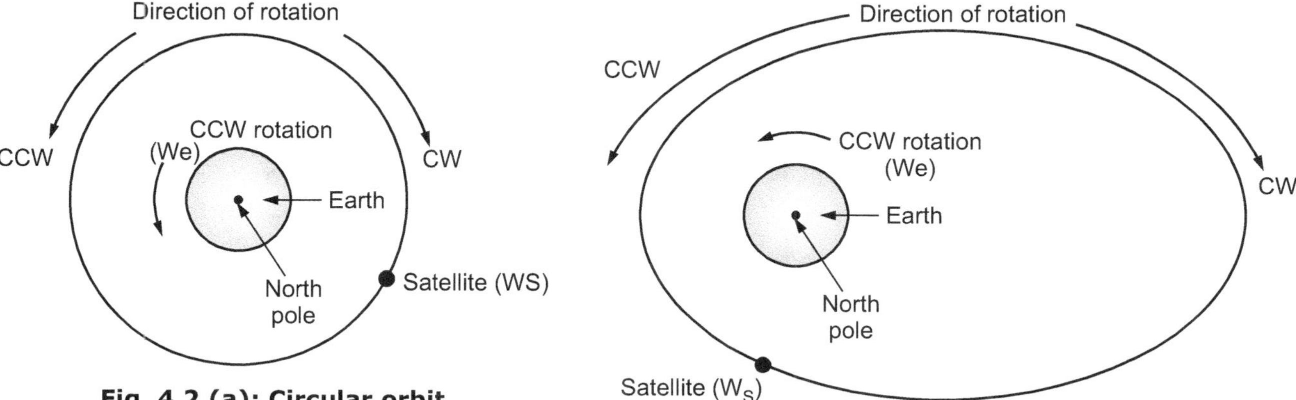

Fig. 4.2 (a): Circular orbit   **Fig. 4.2 (b): Elliptical orbit**

Note: CCW = Counterclockwise
      CW = Clockwise
      ω_e = Angular velocity of earth
      ω_s = Angular velocity of satellite

- *The satellite launched in circular around the earth is known as circular orbit.*
- In a **circular orbit, the speed of rotation is constant**.
- However, in elliptical orbit the speed of rotation depends on the height (the satellite is above earth).
- The speed of the satellite is greater when it is close to the earth than it is farther away.
- If the satellite is orbiting in the same direction as earth's rotation (counter clockwise) and at an angular velocity greater than that of earth ($\omega_s > \omega_e$), the orbit is called **prograde** or posigrade orbit.
- If the satellite is orbiting in the opposite direction as earth's rotation or in the same direction with an angular velocity less than that of earth ($\omega_s < \omega_e$), the orbit is called **retrograde orbit**.
- Most non-synchronous satellites revolve around earth in a **prograde orbit**. Therefore, the position of satellite in non-synchronous orbit is continuously changing in respect to a fixed position on earth.
- **Advantage** of orbital satellites is that propulsion rockets are not required on board the satellites to keep them in respective orbits.
- **Disadvantage** of orbital satellites is the need for complicated and expensive tracking equipment at the earth stations so they can locate the satellite as it comes into view on each orbit and then lock its antenna onto the satellite and track it as it passes overhead.

4.4 VARIOUS COMPONENTS OF SATELLITE SYSTEM

1. Earth Segment:

- The earth segment of a satellite communications system consists of the transmit and receive earth stations. The simplest of these are the home TV receive-only (TVRO) systems, and the most complex are the thermal stations used for international communication networks.

- Also earth segment include the stations, which are on ships at sea, commercial and military and aeronautical mobile stations.

- The earth stations that are used for logistic support of satellites, such as providing telemetry, tracking and command (TT & C) functions, are considered as part of space segment.

2. Space Segment:

- The space segment will obviously include the satellites, but it also includes the ground facilities needed to keep the satellite operational, these being referred to as the tracking, telemetry and command (TT&C) facilities.

- In communication satellite, the equipment which provides the connecting link between satellite's transmit and receive antennas is referred to as the **transponder**.

- The transponder forms one of the main section of the payload, the other is antenna subsystem.

3. Geosynchronous Satellite:

- A geosynchronous satellite is a satellite that remains in geosynchronous orbit around planet, meaning that its orbital period is same as that of earth.

- The orbit, where geosynchronous satellites resolve are known as geosynchronous orbits.

4. Geostationary Satellite:

- The orbit in which satellite completes one revolution around earth in 24 hours, the orbit is called as synchronous or geostationary orbit.

- The satellite that uses synchronous orbit is called geostationary satellite.

Advantages:

(i) In synchronous orbit, satellite angular velocity is same as the earth and so it appears to be stationary.

(ii) It can cover the entire earth's surface.

(iii) It is not necessary to rotate the dish antenna on earth and then tract to satellite.

(iv) Continuous communication is possible.

5. Active Satellites:

- The active satellite carries antenna sysem transmitter receiver and power supply. It works as an active microwave repeaters. Active communication satellite arises from the fact that where communication through relay or an active electronic system.

- The communication capability of active systems with directional antennas rapidly becomes much greater than that of the passive system as the altitude is increased. The principle interest in the society is on systems that are capable of providing appreciable communication capability and hence the modem communication satellites are active satellite systems.

- Now-a-days, space qualified reliable, long-life electronic equipment are available and these have enhanced the capability of active satellite system.

6. Passive Satellites:

- The passive satellite is nothing but a metalized baloon or metallic sphere as is used as passive reflectors. It simply reflects back the signals from one region of earth to the other region. Signal reflectors without any amplification were used before the technology and rocket power were available to provide all the needed subfunctions in the package that was light enough to lift into orbit. Moon bounce experiment in the 1950's showed that signal broadcast from earth could be reflected from the moon and back to another earth station.

- The moon bounce method required tremendous power at the transmitter, since most of the signal power was lost in the round trip distance of nearly half a million miles, and the moon is a poor reflector of radio frequency signals. Also this scheme worked only when the moon was in sight and it could support only a few ground stations since the received signal footprint covered most of the earth even if the transmitted antenna beamwidth was very narrow. Finally, the uplink and downlink frequencies were the same, so full duplex communication were impractical unless each direction used a significantly different frequency from the other.

- Another experiment in the 1960's used a large sphere as the satellite (Echo I) whose surface was covered with metallized Mylar to improve its reflection. Since this satellite was in a much lower orbit than the moon and was a better reflector, it provide better received SNR and smaller footprint. However, this simple passive satellite still provided no gain, had no directional antennas and suffered orbital decay since there were no rocket thrusters to correct any orbit disturbances or errors. Development of more powerful rockets combined with advances in high performance, low power light weight electronics have eliminated passive satellites and have completely replaced them with active satellites.

7. Synchronous Satellites:

- The synchronous satellites also called the geostationary satellites go round the globe in 24 hours so they seem to us on earth to keep station some 36,000 km vertically above the same point on the earth's surface. In fact a stationary satellite is a synchronous satellite that has sufficiently small values of orbital eccentricity and inclination to the equator that the changes in its apparent direction relative to the rotating earth are negligible for practical purposes.

- The other advantages of synchronous altitude is that it is well above the high intensity in radiation belt and above the most intense region of the considerably milder outer belt. Thus, adequate protection can be obtained with negligible weight. The stationary orbit is sunlit over 90% of the time simplifying the generation and storage of power and reducing the number of temperature cycles.

- The strength of the earth's magnetic field is only a hundredth of its value at lower altitudes and spin-dampling time constants are consequently 10,000 times as long as enough to be neglected. The high altitude of the synchronous orbit makes such a satellite visible from 40% of the earth's surface. A single satellite may provide a continuous coverage for the major portion of the world's communication market.

- The ground stations may be of sufficiently low cost that each country served by the system can readily afford to its own terminals and the rf spectrum required for the system can be readily shared with other services. There are few problems with these synchronous communication satellites namely the time delay in connection with telephony and the establishment of the synchronous orbit.

- The synchronous satellites used for communication are widely called the communication satellites. These satellites are classified in terms of their terrestorial coverages for example, global regional or national or in terms of the type of service offered for example, fixed, mobile, maritime, aeronautical etc. or point to point, broadcasting, commercial, military, armateur, experimental etc.

4.5 COMMUNICATION LINK - UPLINK AND DOWNLINK FREQUENCY   (W-15, 16)

- We learnt that the line of sight distance between two antennas was limited by the curvature of the earth and the horizon distance visible from the antenna height.

- Although, higher antennas are the obvious way to increase the achievable distance, but high antenna towers are expensive and impractical in many locations.

- There is an alternative **to use a space satellite as relay station for signals** to overcome the limitations imposed by the earth's horizon.

- A **communication satellite** is designed to receive a signal from a transmitting station on the ground and retransmit it to a receiving station located elsewhere.

- Intermediate link between two earth based stations is known as **communication link**.

- This communication link is known as **up-link** and **down-link**.

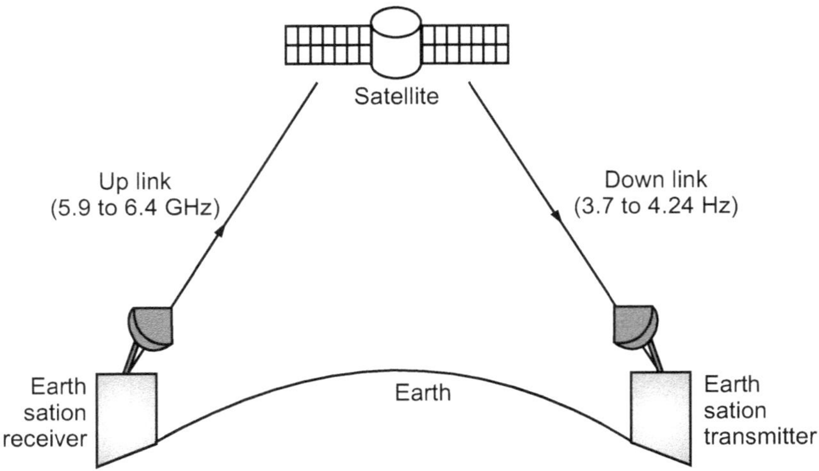

Fig. 4.3: Communication Link

- Satellite is mid-device between uplink and downlink so it is known as transponder.

1. **Uplink:**

- The communication between earth station transmitter towards satellite is known as **uplink**.

    Uplink of a satellite is the frequency at which the earth station is transmitting the signal and satellite receiving it.

> Uplink frequency range = 5.9 GHz to 6.4 GHz.

2. **Downlink:**

- The communication between satellite towards earth station receiver is known as **downlink**.

- **Downlink:**

    Downlink frequency of a satellite is the frequency at which the satellite is transmitting the signal and earth station receiving it.

> Downlink frequency range = 3.7 GHz to 4.2 GHz

3. The Uplink Model:

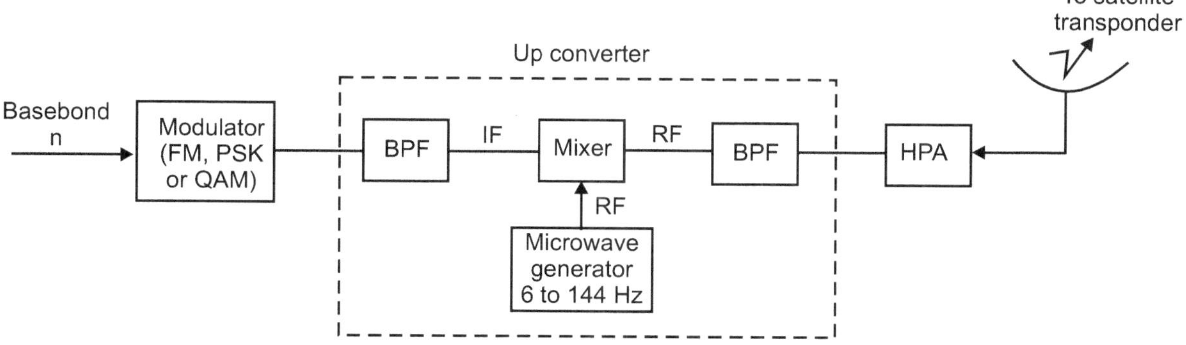

Fig. 4.4: Satellite Earth Station Transmitter Uplink Model

Functions of each block:

IF Modulator:

- The IF modulator convert the input baseband signal to either an **FM**, **PSK** or **QAM** modulated intermediate frequency.

Microwave Generator:

- Generates frequency of 6 to 14 GHz.

Up-converter:

- Mixer and BPF converts the IF to an appropriate RF carrier frequency.

HPA:

- High power amplifier provides adequate input sensitivity and output power to propagate the signal to the satellite transponder.
- The HPA's commonly used are **Klystron's** and **TNT's**.

The Downlink Model:

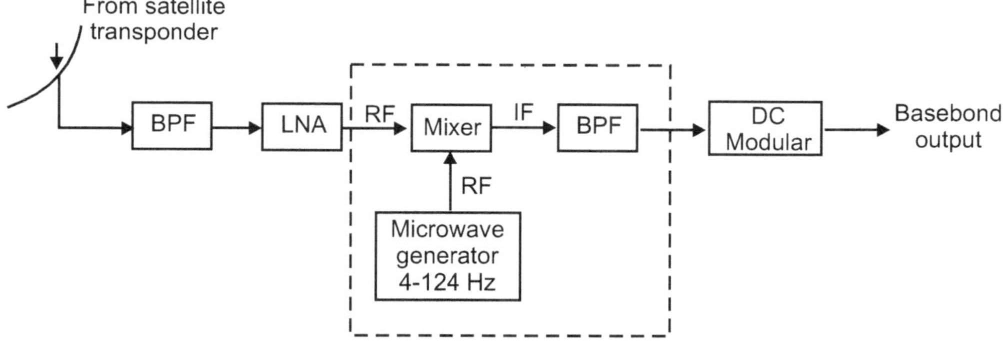

Fig. 4.5: Downlink Model

Functions of each block:

BPF:

- The BPF limits the input noise power to LNA.

LNA:

- The low noise amplifier is a highly sensitive, low noise device.

Mixer-generator-BPF:

- This combination converts the received RF frequency to IF frequency.

Demodulator:

- The most common frequencies used for satellite communication are **6/4** and **14/12 GHz** bands.

- The first number indicates the uplink (earth station to transponder) frequency and second number is downlink (transponder to earth station) frequency.
- Since **'C' bands** is most widely used, this band is becoming overcrowded.
- A typical C band transponder can carry 12 channels, each with a bandwidth of 36 MHz.

4.5.1 Satellite Speed and Period

- The speed of satellite is measured in either **miles per hour, kilometer per hour**, knots.
- The speed varies depending upon the distance of satellite from the earth.
- For a circular orbit the speed is constant, but for elliptical orbit speed varies depending upon the height.
- Very high satellite like communications satellite that are approximately 22,300 miles out revolve much slower, at a speed in the neighbourhood of 6,800 mile/hour.

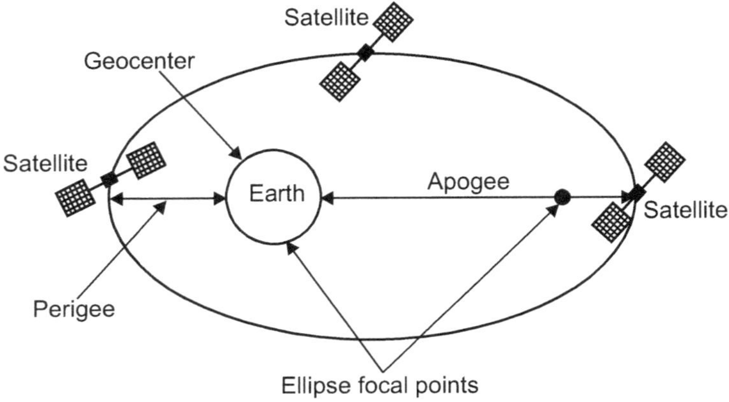

Fig. 4.6: Elliptical Orbit Showing Apogee Perigee

As shown in Fig. 4.6,

- **Apogee** is the point farthest from the earth's surface.
- **Perigee** is the point nearest from the earth's surface.

4.6 LOOK ANGLES (S-15)

- To optimize the performance of a satellite communication system, the direction of maximum gain of an earth station antenna must be pointed directly at the satellite. To ensure that the earth station antenna is aligned, two angles must be determined.

- **Azimuth angle and elevation angle are jointly referred to as the antenna look angle.**

- Elevation angle and azimuth angle both depend on the latitude of the earth station and the longitude of both the earth station and orbiting satellite.

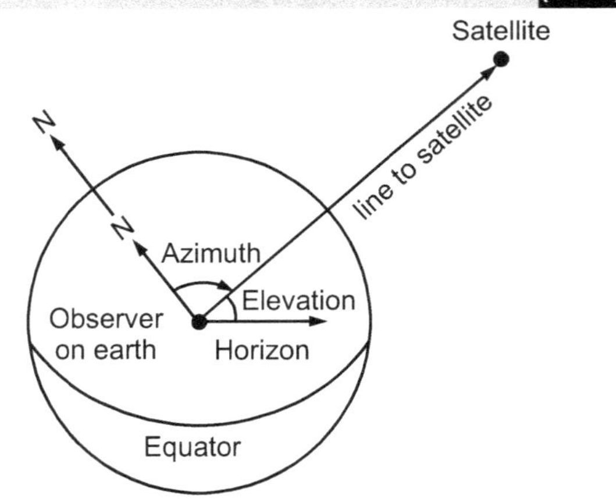

Fig. 4.7: Elevation and Azimuth Angles of Satellite

4.6.1 Elevation Angle (S-15; W-16)

Refer Fig. 4.7.

Definition:

Elevation angle is the vertical angle formed between the direction of travel of an electromagnetic wave radiated from an earth station antenna pointing directly towards a satellite and the horizontal plane.

- The smaller the elevation angle, the greater the distance a propagated wave must pass through earth's atmosphere.

- As with any wave propagated through earth's atmosphere, it suffers absorption and may be severely contaminated by noise.

- If the angle of elevation is too small, and the wave that travels through earth's atmosphere is too long, the wave degrades the transmission quality.

- Generally, minimum acceptable elevation angle is 5°.

4.6.2 Azimuth Angle (W-16)

- Azimuth is the horizontal angular distance from a reference direction, either the southern or northern most point of the horizon.

Azimuth angle is given as,

$$A = 180 + \tan^{-1}\left[\frac{\tan\theta}{\tan\phi}\right]$$

where,   A = Azimuth angle in degrees

θ = S − N

S = Satellite longitude in degrees

N = Earth site longitude in degrees

ϕ = Site latitude in degrees.

- Once the azimuth and elevation angles are known the earth station antenna can be pointed in that direction.

Definition:

- **Azimuth angle is defined as the horizontal pointing angle of the earth station antenna.**

- For navigation purposes, azimuth angle is usually measured in a clockwise direction in degrees from the true north.

- However, for satellite earth stations in the northern hemisphere and satellite vehicles in geosynchronous orbits, azimuth angle is generally referenced to true south (i.e. 180°).

Elevation angle between a straight line from satellite to an earth site and a line tangent to the earth at that site. It is represented by E.

$$\tan E = \frac{\cos\theta\cos\phi - \dfrac{R_e}{R_e + R_o}}{\sqrt{1 - \cos 2\theta \cdot \cos 2\phi}}$$

where,   E = Elevation angle in degrees

θ = S − N

S = Satellite longitude in degrees

N = Earth site longitude in degrees

ϕ = Earth site latitude in degrees

R_e = Earth radius = 6378 km

R_o = Distance of satellite from earth = 35786 km

4.7 ALTITUDE (S-15; W-15)

The highest position from earth is called Altitude.

- The satellite orbit around the earth can be relatively low 50 to several hundred miles or as high as 23000 miles.

- The time for one complete orbit of the satellite is determined primarily by this height.
- A satellite at 100 miles takes 90 minutes for an orbit while one at 10,000 miles takes approximately 12 hours for one orbit, in contrast an orbit as 22,300 miles takes 24 hours.
- The 24 hours orbit when combined with a specific elevation angle is geostationary or geosynchronous where the satellite appears to be stationary over one point on the earth's surface since it and the earth rotate at the same rate.

4.8 FOOT-PRINT   (S-16)

- The area on the earth covered by a satellite depends on the location of the satellite in its orbit, its carrier frequency and the gain of its antenna.
- Satellite engineers select the antenna and carrier frequency for a particular spacecraft to concentrate the limited transmitted power on a specific area of earth's surface.
- The geographical representation of a satellite antenna's radiation pattern is called **footprint** or **footprint map**.

Definition:

The footprint of a satellite is the earth area that the satellite can receive from and transmit to.

- The shape of the satellite footprint depends on the satellite orbital path, height and the type of antenna used.
- The higher the satellite, the more of the earth's surface it can cover.

4.9 FREQUENCY BANDS USED IN SATELLITE COMMUNICATION   (S-15)

QUESTION
1. Explain uplink and downlink frequency bands used in satellite communication.   **(4M)**

- The frequencies reserved for satellite microwave communication are in gigahertz (GHz) range.
- Each satellite sends and receives over two different bands.
- Transmission from earth to the satellite is called up-link.
- Transmission from satellite to the earth is called downlink.
- Table 4.2 gives the band names and frequency ranges.

Table 4.2: Satellite Frequency Bands

Sr. No.	Frequency band	Uplink frequency GHz	Downlink frequency GHz	B.E. GHz	Application
1.	UHF	0.292-0.312	0.25-0.27	0.02	Military
2.	S-band	3.2-3.7	1.8-2.3	0.5	TV transmission used by Doordarshan
3.	C-band	5.9-6.4	3.7-4.2	0.5	TV broadcast. For example, TV programs
4.	X-band	7.9-8.4	7.25-7.75	0.5	Military use, Mobile radio easy
5.	k_u band	14-14.5	11.7-12.2	0.5	TV broadcast, Non-military application.
6.	k_u band (commercial)	27-30	17-20	3	Commercial broadcasting
7.	k_u band (military)	30-31	20-21	1	Military
8	V-band	50-51	40-41	1	Non-military application

4.10 KEPLER'S LAW

- Johannes Kepler was able to derive empirically three laws describing planetary motion.

- Later in, 1665, Sir Isacc Newton (1642-1727) derived Kepler's law from his own laws of mechanics and developed the theory of gravitation.

- Kepler's law apply quite generally to any two bodies in space which interface through gravitation.

- There are three laws of Kepler as follows:

4.10.1 Kepler's First Law

- **"Kepler's first law states that the path followed by a satellite around the primary will be an ellipse".**

- An ellipse has two focal points shown as F_1 and F_2.

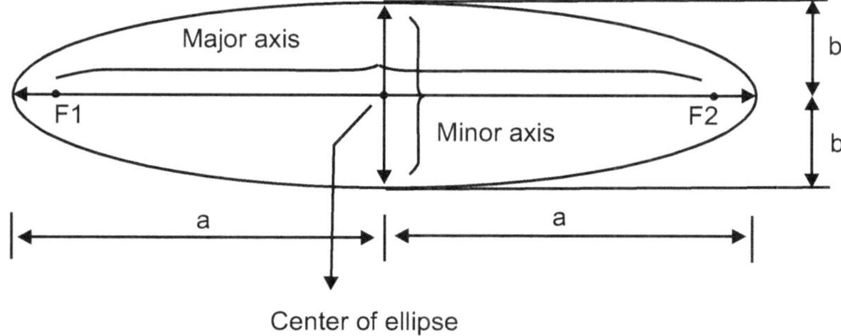

Fig. 4.8: The Foci F_1 and F_2 Semimajor Axis a and Semiminor Axis b of an Ellipse

- The semimajor axis of ellipse is denoted by '**a**' and semiminor axis by "b".

- The eccentricity 'e' is given by

$$e = \frac{\sqrt{a^2 - b^2}}{a} \qquad \qquad \dots (1)$$

4.10.2 Kepler's Second Law

- Kepler's second law states that,

  "For equal time intervals, a satellite will sweep out equal areas in its orbital plane, focused at the barycenter".

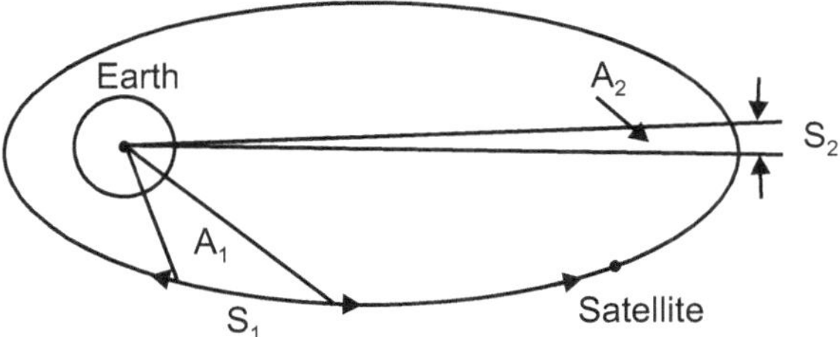

Fig. 4.9: The Areas A_1 and A_2 Swept Out in Unit Time are Equal

4.10.3 Kepler's Third Law

- It state's that, **"The square of the periodic time of orbit is proportional to the cube of mean distance between the two bodies".**

- The mean distance is equal to the semimajor axis "**a**". It can be written as,

$$a^3 = \frac{\mu}{n^2} \qquad \qquad \dots (2)$$

where,

n = Mean motion of satellite in rad/sec.

μ = Earth's geocentric gravitational constant

μ = $3.986005 \times 10^4 \text{ m}^3/\text{s}^2 \qquad \qquad \dots (3)$

4.10.4 Apogee and Perigee Heights

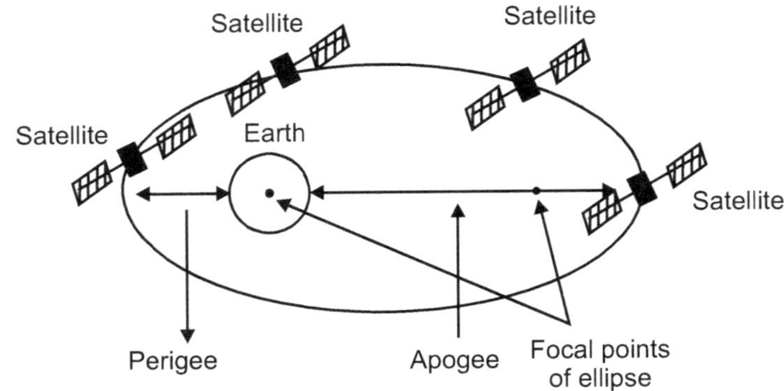

Fig. 4.10: Earth Orbitting Satellite

Apogee Height:

- Apogee height means the furthest distance the moon or a satellite gets from earth in its orbit.
- Apogee '**A**' is related to the semi-major axis and eccentricity as,

$$A = a (1 + e)$$ ... (4)

Perigee Height:

- Perigee means, the closest distance, the moon or satellite gets to earth in its orbit.
- Perigee '**P**' is related to the semi-major axis and eccentricity.

$$P = a (1 - e)$$ ... (5)

4.11 ORBIT PERTURBATIONS

4.11.1 Effects of a Non-spherical Earth

- For a spherical earth of uniform mass, Kepler's third law (equation 2) gives nominal mean motion n_o as,

$$n_o = \sqrt{\frac{\mu}{a^3}}$$ ... (6)

- The o subscript is included as a reminder that this result applies for a perfectly spherical earth of uniform mass.
- However, it is known that, the earth is not perfectly spherical, there being on equatorial bulge and a flattening at the poles, a shape described as an "oblate spheroid".
- When earth's oblateness is taken into account, the mean motion, denoted in this case by symbol n, is modified to

$$n = n_o \left[1 + \frac{K_1 (1 - 1.5 \sin^2 i)}{a^2 (1 - e^2)^{1.5}} \right]$$ ... (7)

where, $\qquad K_1$ = Constant

- the anomalistic period is given as,

$$P_A = \frac{2\pi}{n} s$$ ... (8)

where, **n** is in rad/sec.

4.11.2 Atmospheric Drag

- For near earth satellites, below about 100 km, the effects of atmospheric drag are significant.
- Because the drag is greatest at the perigee, the drag acts to reduce the velocity at this point, with the result that satellite does not reach the same apogee height on successive revolutions.
- The result is that the semimajor axis and the eccentricity are both reduced.

- Drag does not noticeably change the other orbital parameters, including perigee height.
- An approximate expression for change of major axis is

$$a \cong a_o \left[\frac{n_o}{n_o + n_o' \, (t - t_o)} \right]^{2/3} \qquad \ldots (9)$$

- where, "o" subscripts denote values at the reference time t_o and n_o' is the first derivative of mean motion.

Important Points

- A communication satellite is a **microwave repeater.**
- The satellite is POSITIONED at about **36,000 kms above the earth,** where it experiences very less gravitational pull of the earth.
- A good quality service can be continuously ensured by a global operational satellite system.
- A satellite system should cater for carrying different types of signals which include telephony, data, telegraphy and facsimile. While television signals can be transmitted separately.

  For example,

  INTELSAT-I, INTELSAT-II, INTELSAT-III, INTELSAT-IV, INTELSAT-IVA.
- An artificial satellite needs to have orbit, the path in which it travel around the earth.
- Most of satellites mentioned thus far are called orbital satellites, which are non-synchronous.
- Satellites which are not stationary with respect to earth are called non-synchronous satellites.
- A satellite remains in orbit because the centrifugal force caused by its rotation around the earth is counter-balanced by earth's gravitational pull.
- Non-synchronous satellites rotate around the earth in an elliptical or circular pattern, so there are two types of orbit.
  1. Circular orbit.
  2. Elliptical orbit
- *Uplink of a satellite is the frequency at which the earth station is transmitting the signal and satellite receiving it.*

  Uplink frequency range = 5.9 GHz to 6.4 GHz.
- Downlink frequency of a satellite is the frequency at which the satellite is transmitting the signal and earth station receiving it.
- The orbit in which satellite completes one revolution around earth in 24 hours, the orbit is called as synchronous or geostationary orbit.
- The speed of satellite is measured in either **miles per hour, kilometer per hour**, knots.
- **Apogee** is the point farthest from the earth's surface.
- **Perigee** is the point nearest from the earth's surface.
- Azimuth angle and elevation angle are jointly referred to as the antenna look angle.
- Elevation angle is the vertical angle formed between the direction of travel of an electromagnetic wave radiated from an earth station antenna pointing directly towards a satellite and the horizontal plane.
- Azimuth angle is defined as the horizontal pointing angle of the earth station antenna.
- The highest position from earth is called Altitude.
- The footprint of a satellite is the earth area that the satellite can receive from and transmit to.

- A communication satellite is a **microwave repeater.**

- The frequency between uplink and downlink prevents the interference between the signals.

- **Uplink** of a satellite is one in which the earth station is transmitting the signal and satellite receiving it.

- **Downlink** of a satellite is one in which the satellite is transmitting the signal and earth station receiving it.

- The **orbit** is a path of satellite at a height above the earth surface when it revolves round the earth.

- A **geo-stationary orbit** is the path of satellite around earth at which the velocity of satellite is so maintained that from earth it appears to be stationary.

- **Apogee** is the point farthest from the earth's surface.

- **Perigee** is the point nearest from the earth's surface.

- The satellite location is specified at a point on surface of earth directly below the satellite, this point is known as **Sub Satellite Point** (SSP).

- The **look angles** are the angles to which as earth station antenna must be pointed to communicate with the geo-synchronous satellite.

- The **angle of elevation** is the angle between the horizontal plane and the pointing the direction of the antenna.

- **Azimuth angle** is defined as the angle by which the antenna pointing at a horizon must be rotated clockwise around its vertical axis with respect to north.

- The satellite down-link antenna transmits signals towards earth in a particular pattern called **foot-print.**

Practice Questions

1. Give the components of satellite system.
2. Explain earth and space segment.
3. Explain active and passive segment.
4. Give idea about geostationary and geosynchronous satellite.
5. Draw the frequency allocation band of satellite.
6. Explain uplink and downlink frequency bands with model.
7. Explain look angle and elevation angle.
8. Draw communication satellite orbits and explain the same.
9. State Kepler's law, apogee and perigee.
10. Explain Kepler's 1st, 2nd and 3rd law.
11. Give effect of a non-spherical earth on satellite.
12. Explain atmospheric drag.
13. Explain station keeping and footprint.

Satellite Segments and Services

Syllabus

5.1 Satellite Earth Station: Block Diagram; Antenna Subsystem, LNA, Power Subsystem, Telemetry Tracking and Command (TTAC) Subsystem, Attitude Control, Spinning Satellite Stabilization, Momentum Wheel Stabilization, Station Keeping, Thermal Control Transponder: Single, Double Conversion and Regenerative Type

5.2 Space Link: Equivalent Isotropic Radiated Power (EIRP), Transmission Losses : Free-space Transmission Loss, Feeder Losses, Antenna Misalignment Losses, Fixed Atmospheric and Ionosphere Losses

5.3 Satellite Applications:
GPS: Global Positioning System (GPS) Concept, Working Principle, Transmitter and Receiver
VSAT: Overview, Architecture, Working Principle, Applications

About this Chapter

At the end of this chapter students will be able to:
- Describe with sketches the functions of the given subsystem of the satellite earth station.
- Describe the given type of control systems associated with the Satellite.
- Describe with sketches given applications.

5.1 INTRODUCTION

- Satellites overcome the difficulties and limitations in broadcasting beyond the visible horizon.
- A satellite system represents a careful and finely tuned integration of electronics, mechanical structure, rocketry and antenna design supported by a system of earth based ground stations, computers and radar for tracking satellite position precisely.
- Different orbits provide a selection of coverage of the earth's surface with orbital times ranging from 90 minutes to 24 hours.
- Using frequencies in gigaheartz range, satellites are effective and predictable in performance but have very large path losses from earth to satellite to earth.
- A complete system requires high gain antennas, sensitive and low noise front and receiver stages and careful planning.

5.2 SATELLITE EARTH STATION BLOCK DIAGRAM

QUESTION

1. Sketch block diagram of satellite earth station and state functions of each block. **(8M)**

Although the communications satellite is the most sophisticated of communication systems in design and construction, the earth station (or ground station) is a critical part of successful satellite communication system.

- Fig. 5.1 shows the block diagram of ground station.

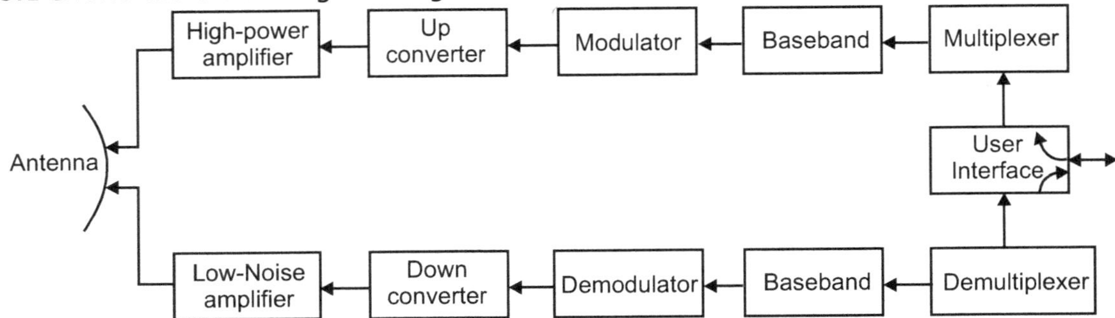

Fig. 5.1: Satellite Earth Station Block Diagram

- A ground station has five main sub-systems: the interface to the earth-based signal source (the user who is sending and receiving a message), the transmit chain, the receive chain, the antenna and the antenna control.

- **Functions of each block:**

- Transmit chain.

- **Multiplexer:**

Transmits signals simultaneously many inputs from transmitter.

- **Baseband:**

Original information signal in audio range.

- **Modulator:**

Baseband signal converted into wave.

- **Up-conversion:**

Modulated signal typically 70 MHz is up converted into the actual broadcast frequency such as 6 GHz.

- **High-power Amplifier:**

Signal power amplified upto 1 MW to reach deep space satellites and is transmitted towards satellite through transmitting antenna.

Receiver:

- **Low-noise Amplifier:**

Provides enough gain while maintaining maximum possible signal to noise ratio.

- **Down converter:**

The received signal down converted to IF again typically 70 MHz.

- **Demodulation:**

Modulated signal is converted back into original baseband signal.

- **Demultiplex:**

Converts one signal into many.

5.3 SATELLITE SUBSYSTEMS   (S-15, 16)

QUESTION
1.   State functions of telemetry and tracking system.   (4M))

* A satellite consists of many subsystem function, all carefully integrated into a single system. These are :

1. Power subsystem.
2. Communication channel subsystem.
3. Attitude control subsystem.
4. Thermal control system.
5. Telemetry tracking and command subsystem.
6. Main and auxiliary propulsion subsystem.
7. Antenna subsystem.

Fig. 5.2 shows the functional block diagram of satellite.

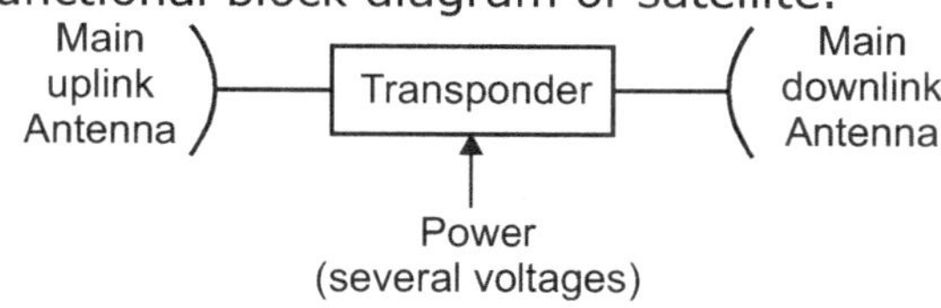

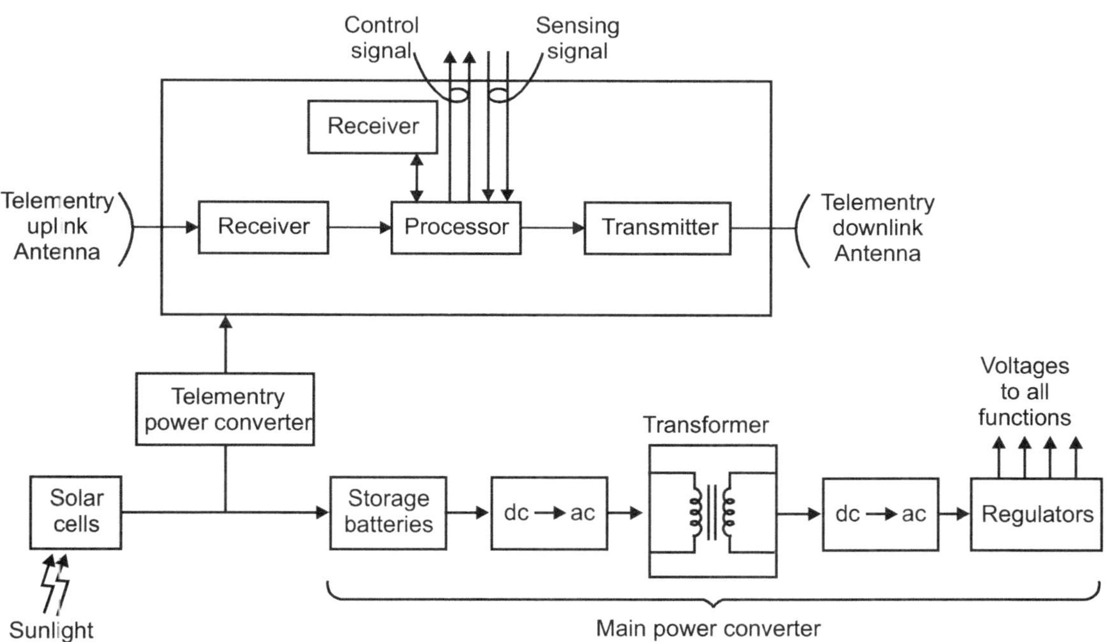

Fig. 5.2: Key Satellite Functional Blocks (S-15)

5.3.1 Power Subsystem

(S-16)

* The power subsystem is responsible for providing the primary dc power and the regulated, secondary supply voltages for the satellite circuits.
* Depending on design and complexity, a communication satellite requires several hundred to several thousand watts of electrical power.
* Primary power originated from solar cells, which converts sunlight into electricity, large panels of cells are required.
* These extend from the satellite like wings, and the cells must be oriented to face the sun directly for maximum output.

- In the night time solar cell power is not used directly, instead illuminated cells charge storage batteries on board, which provide the primary power to the satellite power converter.

- The storage batteries must be able to provide for more than 12 hours of operation without recharging.

- The internal circuitary of the satellite requires a variety of dc voltages from +5 and ± 15V for low level logic and amplifiers, to several hundred volts for power amplifier.

5.3.2 Communication Channel Subsystem 

- The central purpose of the satellite is the communications channel subsystem – everything else exists to serve it.

- A transponder takes the received signal from the antenna at the uplink frequency, heterodynes (mixes) it to the downlink frequency and amplifiers it before transmitting.

- A complete basic transponder consists of an input filter, low noise input amplifier, bandpass filter, mixer plus local oscillator, another bandpass filter and the power amplifier.

- This type of transponder is open and transparent and handles any signal that fits into it bandwidth.

- The output signal format and modulation are identical to those of input signal and only the carrier frequency is changed.

5.3.3 Attitude Control Subsystem

Attitude Control:

- In addition to maintaining the position of the satellite in orbit, some means must be provided to position the satellite for optimum performance, this is called attitude control.

- The attitude of the satellite must be controlled so that the antennas can be pointed towards the correct locations on earth.

- Attitude control is also necessary in some satellites to keep the satellite's solar panels pointed towards the sun so that maximum power is produced at all times.

- In some application, it may be necessary where selected sensors or a TV camera must be pointed in the right direction.

- Attitude control is maintained by a combination of satellite stabilization techniques and jet thrusters for correctional purposes.

- The attitude of the satellite is first determined right after the satellite is put into stable orbit.

- At some point, the various jet thrusters on the satellite are actuated to move the satellite in such a way that the correct attitude is assumed.

- Once the initial attitude of the satellite is set, it must be maintained in this position. This is done by spin stabilization method.

5.3.4 Attitude Control Subsystem

Control Subsystem:

- All satellites contain an attitude stabilization subsystem with built-in jet thrusters for making corrections to the attitude of the satellite.

- By way of the command and control subsystem and the on-board computer, the attitude stabilization subsystem receives signals that fire the appropriate thrusters at the proper time to achieve an adjustment in attitude.

- This is to reorient antennas, maximize exposure of the solar panels to the sun, or make corrections indicated by photo or infrared sensors that provide orientation input for a three axis stabilized satellite.

- The attitude stabilization system may also consist of gyroscopes, flywheels for spin stabilization and antenna de-spin mechanisms.

- In addition to jet thrusters for attitude stabilization, most satellite also contain a propulsion system.

- A propulsion system is the apogee kick motor used to put the satellite into final orbit.

- The propulsion system is also operated by the on-board computer in response to the command control subsystem.

5.3.5 Telemetry Tracking and Command Subsystems (TTAC)   

QUESTION
1. Describe the function of telemetry and tracking in satellite communication system.   **(4M)**

- Telemetry tracking and command subsystems provide a way for the ground station to manage the satellite.

- The communications channels used for this function are entirely separate from the communications channels used to replay user messages from uplink to downlink via status telemetry, the satellite transmits information back to the ground station about hundreds of vital internal voltages, currents, temperatures and fuel levels as well as positional and orientation information.

- This information is used to monitor 'health' and operation continuously indicating both sudden failures and signs of potential problems.

- There is also command telemetry for the ground station to command the satellite to activate its rocket thrusters to change position or orientation to cause the electronics to switch to a different mode of operation (modulation, frequency, number of channels and bandwidth) or to electronically reconfigure same of the internal interconnections.

- The telemetry and control subsystem is so critical to ensuring that the satellite achieves proper orbit and maintains and corrects orbital position as well as for monitoring the performance and even fixing the satellite, the telemetry plus control electronics are as separate as possible from the communication electronics.

- The telemetry and control subsystem has its own battery, power converter, uplink, downlink and even antenna system.

5.3.6 Main and Auxiliary Propulsion Subsystem

- The satellite is launched into its approximate desired orbit by a large rocket which may be carrying several satellite at the same time.

- The satellite then uses its own main propulsion system to go to its final orbit, correct for any unavoidable errors in the initial launch.

- Since only a limited amount of main rocket fuel is available on board the satellite, this orbit adjustment must be correct on the first or second try.

- Some surveillancy (spy) satellites have additional main propulsion fuel so that they can actually shift from one orbit to an other via ground command, to take pictures of other parts of the earth or listen into signals from different areas but this naturally limits their overall useful life.

5.3.7 Antenna System

- All satellites have one or more antennas used for receiving signals from the earth (ground) station and transmitting information back to earth.

- Antenna subsystem have a dual function.

- Numerous antennas are generally required because of various satellite requirements.

- For example, one antenna may be used for communications relay purposes and another antenna may be dedicated to the telemetry, tracking and control functions.

- Most of these are highly directional gain antennas that must be accurately pointed.

- However, most satellites contain an omnidirectional antenna that is used on the command receiver so that the satellite may pick-up ground (or earth) control signals during launch, orbit transfer and other periods prior to full attitude stabilization.

5.3.8 The Satellite Transponder

- A transponder is a part of satellite, which is a combination of transmitter and receiver.

- The main function of the transponder is **frequency translation** and **amplification**.

- Based on frequency translation process, there are three basic transponder configurations.

  (i) Single conversion transponder   (ii) Double conversion transponder

  (iii) Regenerative transponder

- To down convert the signal frequency and to increase of signal the transponder is required.

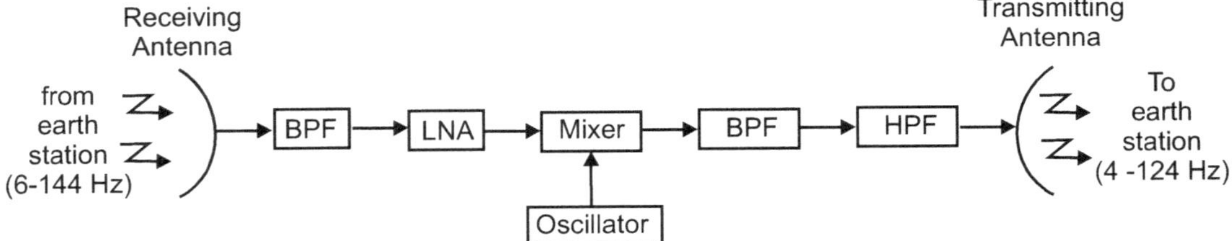

Fig. 5.3: Block Diagram of Satellite Transponder

Block Diagram Description:

Receiving Antenna:

- The uplink signal is received by the receiving antenna.

BPF:

- The received signal is first band limited by Band Pass Filter (BPF).

LNA:

- The signal is amplified by low noise amplifier.

Mixer-oscillator BPF:

- The amplified signal is then frequency translated by a mixer and oscillator. Here, only the frequency is translated from high band up-link frequency to downlink frequency.

HPA:

- The down converted frequency signal amplified by high power frequency.

Transmitting Antenna:

- The down link signal is transmitted to receiver earth station through power transmitting antenna.

NPA:

- The down converted frequency signal amplified by high power frequency.

Transmitting Antenna:

- The downlink signal is transmitted to receiver earth station through power transmitting antenna.

- Satellite transponder is having two types:

5.3.8.1 Single-conversion Transponder

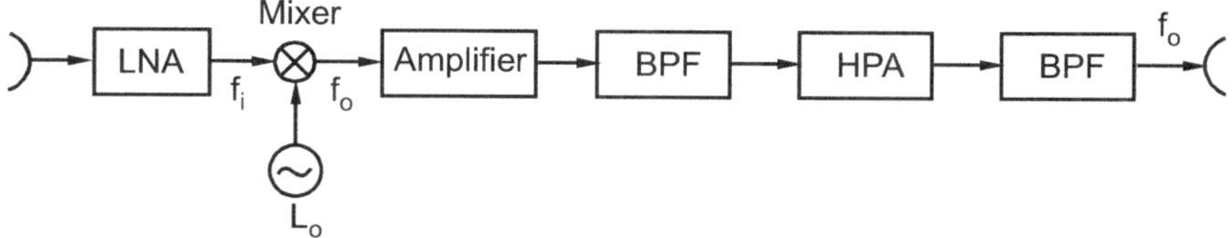

Fig. 5.4

Description:

* In this transponder only a single-frequency translation process takes place.
* First uplink frequency signal is picked-up by the receiving antenna and is routed to LNA (Low Noise Amplifier).
* The signal is very weak at this point, so LNA amplifies the signal
* Once the signal is amplified, it is translated in correct frequency by mixer.
* The output of mixer is then amplified again and fed to band pass filter (BPF1).
* BPF1 allows only a desired down-link signal of 4 GHz.
* At last, the down-link signal is amplified by high power amplifier (HPA) usually TWT (Travelling wave tube).
* Again output of BPF2 is fed to the down-link antenna.
* If common antenna is used for transmission or reception then diplexer is used to share the antenna.

5.3.8.2 Double-conversion Transponder

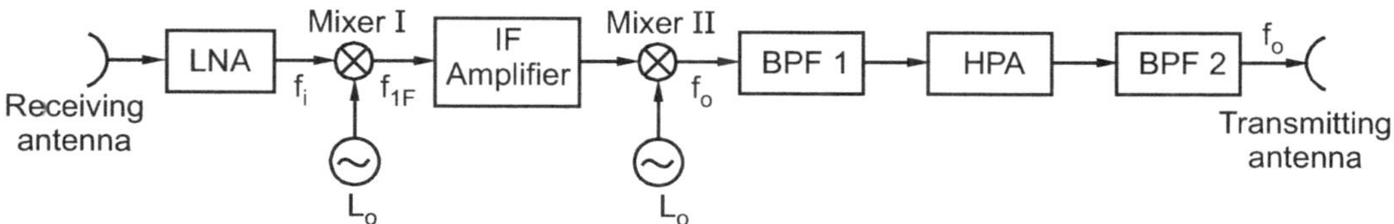

Fig. 5.5

* First uplink signal is received by the receiving antenna.
* LNA amplified the received signal.
* Amplified signal first fed to first mixer (1).
* The mixer 1 translates the received signal frequency into intermediate frequency (typically 70 and 150 MHz). If output is fed to an IF amplifier.
* The output of IF amplifier is fed to another mixer 2.
* The mixer 2 translates the signal to the output frequency.
* BPF1 filters the output signal and eliminates the unwanted output.
* HPA increases the output signal level.
* Again output signal is passed through BPF2 to filter out the harmonics etc.
* At last, transmitting antenna sends the signal over the down link.
* This transponder provides greater flexibility in filtering and amplification.

## 5.3.9 Sation Keeping					(S-15, 16)

* The process of firing the rockets underground control to maintain or adjust the orbit is referred to as sation keeping.
* Once a satellite is in orbit, the forces acting on it tend to keep it in place.
* If the satellite speed and height during launch are accurately controlled, the satellite will enter the proper orbit and remain there.

- However, even when a very accurate launch, the satellite will drift somewhat in its orbit.
- The drift is particularly undesirable in a geosynchronous satellite whose position is supposed to remain fixed for reliable continuous communication.
- Orbital drift is caused by a variety of forces.
   1. The gravitational pull of the sun and the moon affect the satellite's position.
   2. The earth's gravitational field is not perfectly consistent at all points on the earth. This is because earth itself is not a perfect sphere, instead earth is little but fatter at equator and is flattened somewhat at poles.
- Because of this drift, the orbit of the satellite must be periodically adjusted. Most satellite contain small rockets or thruster jet for that purpose.
- These rockets placed at various positions on the various positions on the satellite, that can be used to speed-up or slow down the satellite for compensating the orbital drift.
- Depending on how accurate the orbit must be, these rockets may be fired as every several weeks or as little as once per year.
- The rockets can be conventional rocket motors used to launch the satellite and put in into orbit.
- Most satellites have several thruster jets to make various satellite position adjustment possible.
- Thus, the process of firing the rockets under ground (or earth) control to maintain or adjust the orbit is referred to as station keeping.

5.4 SPACE LINK

- Link power budget calculations basically relate two quantities the transmit power and the receiver power.
- Link-budget calculations are usually made using **decibel** or **decilog** quantities.

5.4.1 Equivalent Isotropic Radiated Power (EIRP)

- EIRP is a measurement of radiated output power from an ideal isotropic antenna in a single direction.
- An isotropic antenna is meant to distribute power equally in all directions.
- When we channel that power into a single direction and calculate the power it is known as **EIRP**. It will be the maximum power emitted by the antenna in the direction with highest antenna gain.
- EIRP is calculated using following formula:

$$\boxed{EIRP = P_T - L_C + G_a} \qquad \text{... (5.1)}$$

where,   EIRP = Output power of a signal when it is concentrated into a smaller area by antenna.

P_T = Output power of the transmitter (dBm).

L_C = Cable loss (dB).

Ga = Antenna gain (dBi).

5.5 TRANSMISSION LOSSES

- Transmission losses are classified as follows :

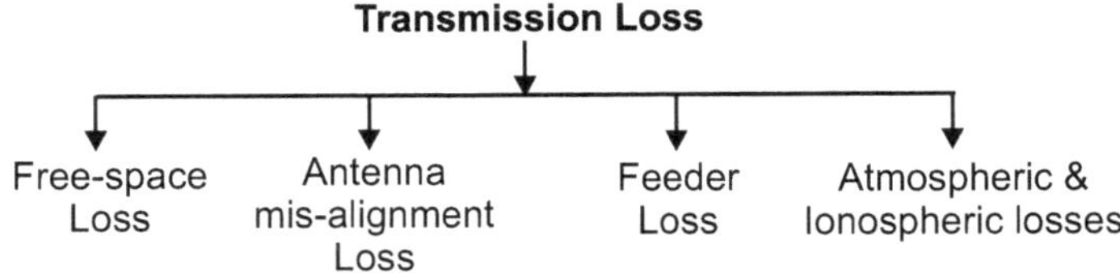

5.5.1 Free-Splace Transmission Loss

- It is used to predict the strength of RF signal at a particular distance, but in real world, there are many obstacles, reflections and losses which need to be accounted for when estimating the signal at a location.

- Free space transmission loss can be given as,

$$[FSL] = 32.4 + 20 \log r + 20 \log f$$

Example: The range between a ground station and a satellite station is 42,000 km, calculate free-space loss at a frequency of 6 GHz.

$$[FSL] = 32.4 + 20 \log 42,000 + 20 \log 6,000$$

$$= \textbf{200.4 dB}$$

5.5.2 Feeder Loss

- Losses will occur in the connection between the receive antenna and the receiver proper.

- Such losses will occur in the connecting waveguides filters and couplers, these will be denoted by RFL or [RFL] dB receiver feeder losses.

5.5.3 Antenna Misalignment Loss

- Antenna measurement technique refers to the testing of antennas to ensure that the antenna meets specifications or simply to characterize it.

- Typical parameters of antennas are gain, radiation pattern, beamwidth, polarization and impedance.

- Antenna misalignment loss is calculated by the formula:

$$L_A = 12 \times \left(\frac{a}{\theta_3 \text{ dB}} \right)^2$$

where,

L_A = Loss (dB)

a = Misalignment (deg)

θ_3 dB = Beamwidth (deg)

5.5.4 Atmospheric and Ionospheric Losses

- This kind of losses derives from the absorption of energy by atmospheric gases. They can assume two different types :

(i) Atmospheric attenuation   (ii) Atmospheric absorption

- The major distinguishing factor between them is their origin. Attenuation is weather related, while absorption comes in clear-sky conditions.

- Likewise, these losses can be due to ionospheric, tropospheric and other local effects.

5.6 SATELLITE APPLICATIONS

- Satellite applications can be broadly classified into the following categories namely –

    1. Satellite communication applications.

    2. Remote sensing and earth observation application.

    3. Meteorological applications.

    4. Military application.

    5. Scientific and technological applications.

- In the field of satellite communication applications, satellite television, telephone and data communication are the major application areas. In the category of remote sensing/earth observation applications the typical ones are discovery of hidden mineral resources, terrain mapping etc. The meteorological applications include weather forecasting, flood forecast, melting of glaciers etc. Military applications include providing strategic communication links between border forces and headquarters spying, providing navigational aids to ships, aircrafts, etc. In the category of satellites for science and technology, the applications include use of satellites for astronomical research, monitoring of different layers of atmosphere etc.

- Some details applications are as follows :

5.6.1 Global Positioning System (GPS)

- GPS system consists of three "segments" called the control segment, the space segment and the user segment.

Control Segment:

- The control segment is responsible for detecting satellites that are not broadcasting properly, or that are not in proper orbit.

Space Segment:

- It is composed of a constellation of satellites orbiting approximately 20,000 km above the earth.
- The full constellation is defined as 24 satellites.

User Segment:

- It is the term given to all receivers listening to the satellites at any time.
- There is no organisation to the user segment, but for any user, it consists of the receiver currently in use and its associated antenna.

Principle of Operation:

- The satellite orbiting above the earth, simply broadcast their location and current time.
- The receivers listen to several satellites and from the broadcasts determine what time it is and where the receivers are located.

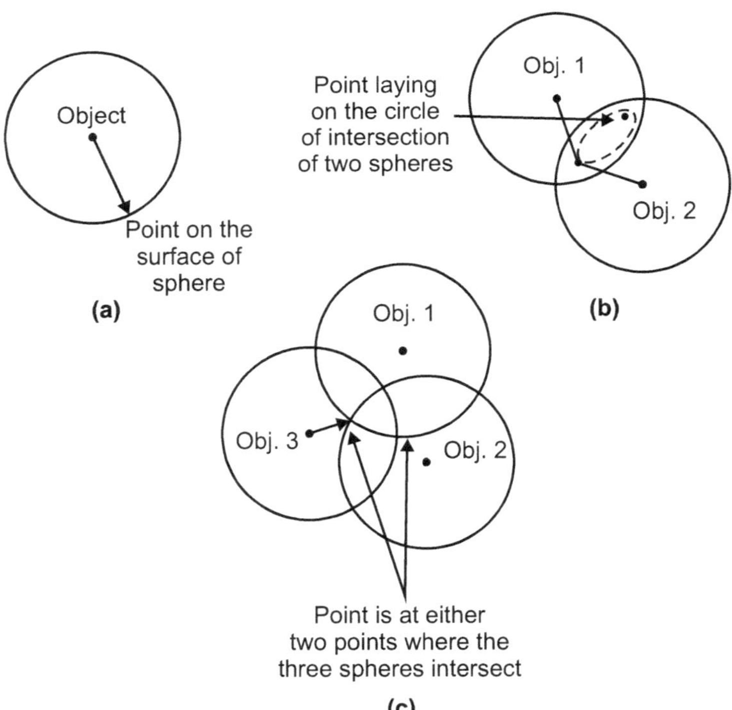

Fig. 5.6: Object Position Determination

5.6.2 Very Small Aperture Terminals (VSATs)

Concept:

- VSATs are used for providing one-way or two-way data broadcasting services, point to point voice services and one-way video broadcasting services.
- VSAT networks are ideal for centralized networks with a central host and a number of geographically dispersed terminals.

Working Principle:

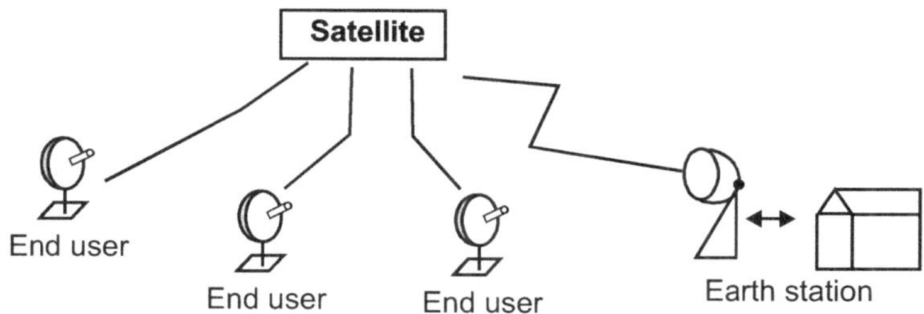

Fig. 5.7: VSAT Communication System

- As shown in Fig. 5.7, there is a **HUB** or **earth station**. It is also called as **central orifice**. This HUB is connected with various end users via. satellite.
- At the earth station, a computer is installed which control the communication of earth station with various end users via. satellite.
- There may be various end users established at different locations. These end users cannot communicate directly with one another.
- If they want to communication between them then a signal goes from end users to HUB via. satellite. Then HUB receive this signal and then re-transmit it to required end users via. satellite.

Advantages of VSAT:

1. No security issues.
2. Cost effective.
3. Good coverage area.
4. Good stability and flexibility.
5. High reliability.

Disadvantages of VSAT:

1. Goes through environmental effects.
2. Connectivity issues.
3. Time delay.

Applications of VSAT:

1. Distance education.
2. PoS transaction, banking, inventory.
3. Corporate networks.
4. High-speed internet access.
5. Financial management.

Important Points

- The process of firing the rockets underground control to maintain or adjust the orbit is referred to as sation keeping.
- A satellite consists of many subsystem function, all carefully integrated into a single system. These are:

1. Power subsystem.
2. Communication channel subsystem.
3. Attitude control subsystem.
4. Thermal control system.

5. Telemetry tracking and command subsystem.

6. Main and auxiliary propulsion subsystem.

7. Antenna subsystem.

- The power subsystem is responsible for providing the primary dc power and the regulated, secondary supply voltages for the satellite circuits.

- The central purpose of the satellite is the communications channel subsystem – everything else exists to serve it.

- In addition to maintaining the position of the satellite in orbit, some means must be provided to position the satellite for optimum performance, this is called attitude control.

- All satellites contain an attitude stabilization subsystem with built-in jet thrusters for making corrections to the attitude of the satellite.

- Telemetry tracking and command subsystems provide a way for the ground station to manage the satellite.

- The satellite is launched into its approximate desired orbit by a large rocket which may be carrying several satellite at the same time.

- All satellites have one or more antennas used for receiving signals from the earth (ground) station and transmitting information back to earth.

- Satellite applications can be broadly classified into the following categories namely –

    1. Satellite communication applications.

    2. Remote sensing and earth observation application.

    3. Meteorological applications.

    4. Military application.

    5. Scientific and technological applications.

Practice Questions

1. Draw the block diagram of satellite earth station and explain the same.
2. Describe antenna subsystem.
3. Explain TTAC and power subsystem.
4. Give idea about station keeping.
5. Draw the block diagram of single conversion transponder.
6. Explain double conversion transponder.
7. Explain EIRP.
8. Give brief idea about transmission losses.
9. explain feeder and antenna misalignment loss.
10. Explain free-space transmission loss.
11. Explain GPS as an satellite application with neat diagram.
12. Explain VSAT with working principle and architecture of the same.

✳✳✳

www.ingramcontent.com/pod-product-compliance
Lightning Source LLC
LaVergne TN
LVHW080612200726
843509LV00007B/288